Microelectronics and microcomputer applications

SELECTED READINGS
FOR ENGINEERING EDUCATORS

Papers from two special issues of the
International Journal of Electrical Engineering Education

Edited by **M. G. Hartley** and **A. Buckley**

MANCHESTER UNIVERSITY PRESS

© Manchester University Press 1983

Whilst copyright in the volume as a whole is vested in Manchester University Press, copyright in individual chapters belongs to the respective author, and no part of any chapter may be reproduced or utilised in any form or by any means, electronic or mechanical, including photocopying, recording, or by any information storage and retrieval system, without the express permission in writing of both author and publisher

Published by
Manchester University Press
Oxford Road, Manchester M13 9PL

British Library Cataloguing in Publication Data

Microelectronics and microcomputer applications.
  1. Microelectronics 2. Microprocessors
I. Hartley, M. G. II. Buckley, A.
621.381'71 TK7874

ISBN 0-7190-0905-7

Printed in Great Britain
by H Charlesworth & Co Ltd, Huddersfield

Microelectronics and
microcomputer applications

CONTENTS

	Foreword	page vii
Part 1	*Design and fabrication of integrated circuits*	
1	Low-cost silicon device fabrication in degree courses M. J. Morant	2
2	A novel university approach to teaching microelectronics K. F. Poole	11
3	A CMOS breadboard gate array for educational and R and D purposes S. L. Hurst	23
4	A practical approach to digital integrated circuit design using uncommitted logic arrays P. J. Hicks	38
5	Production of microelectronic components at UMIST H. D. McKell	53
6	Microelectronics teaching at the University of Canterbury, New Zealand L. N. M. Edward	59
7	Linear integrated circuit design in the curriculum H. E. Hanrahan and S. J. West	70
8	Liquid crystal displays as a topic for undergraduate laboratory work B. Lawrenson	75
Part 2	*Available components*	
9	Recent developments in the design of microcomputer system components P. G. Depledge	87
Part 3	*Laboratory use of microcomputers*	
10	A single-bit microcomputer experiment R. M. Hodgson and E. J. Hamilton	104
11	Analog to digital conversion using a microprocessor A. M. Chadwick and W. Herdman	109
12	A microcomputer-controlled experiment to measure semiconductor material properties K. E. Singer and H. D. McKell	116
13	A microcomputer-assisted power system laboratory G. G. Richards and P. T. Huckabee	125

Part 4	*Computer-aided design*	
14	Developments in computer-aided circuit design D. Boardman and *I. R. Ibbitson*	130
15	LINSIM — a linear electrical network simulation and optimisation program *L. N. M. Edward*	138
16	Computer-aided design of digital systems at functional level *L. M. Patnaik* and *U. Dixit*	156
Part 5	*Available popular small computers*	
17	A review of some popular small computers *M. J. Bosman*	166
Part 6	*Computer-aided learning (CAL)*	
18	Efficient computer-based teaching through task syndication *C. McCorkell* and *R. N. Wilson*	188
19	Teaching computer architectural concepts using interactive digital system simulation *A. J. Walker*	195
Part 7	*New courses and teaching methods*	
20	An undergraduate course in real-time computer systems *H. S. Bradlow*	210
21	An advanced electronics teaching laboratory *R. M. F. Goodman, G. E. Taylor* and *A. F. T. Winfield*	221
22	Microprocessor engineering and digital electronics: an example of an M.Sc. course at UMIST *M. G. Hartley*	231

FOREWORD

In the late 1940s, a team of engineers working at the Bell Telephone Laboratories discovered, almost accidentally, a physical effect which provided the basis for the first transistors. Thus was born the age of microelectronics and, over thirty years later, the consequent revolution is still in full flood.

By the middle of the following decade the thermionic valve had been supplanted by semiconductor devices for most low power and medium frequency applications. Five years later the great majority of applications throughout the power and frequency spectra had become the exclusive province of the new technology.

For electronic engineers these were exciting times: the new devices, with their small physical size, higher reliability and reduced power consumption, created new freedoms for circuit design. New devices with improved characteristics were announced weekly and put to immediate use. Instruments, computers and other electronic products, whose reliability and size would have been impossible to achieve using valves, were devised and applied.

In 1960, the electronic engineer's main preoccupation was still the creation of circuits from primary component elements — resistors, capacitors, transistors and inductors. Advanced forms of traditional 'valve' circuits such as monostables, bistables, linear amplifiers and oscillators were translated to and improved by the application of semiconductor technology. Soon, however, the first packaged circuits appeared, mainly in the field of digital logic. Albeit that these were encapsulated assemblies of discrete components, the circuit designer now found himself presented with a kit of circuit sub-assemblies — gates, inverters and bistables — with compatible interfaces and guaranteed performances.

One result of these advances was to allow the electronic engineer to escape from the intimate details of circuit design and to concentrate on the architecture of circuits and systems. The packaged component manufacturers assumed responsibility for circuit design and the application engineer began to concentrate on how these sub-assemblies could best be used. As a consequence, there emerged a great activity in the investigation and devising of major circuit elements such as digital adders, multipliers, peripheral drivers and analogue amplifiers, tuning circuits and waveform generators.

The next major step forward came with the integration of these circuits onto a reducing number of semiconductor substrates. Although this new technique often simplified the circuit details to the extent of reducing the overall performance, the reductions in size and increases in reliability achieved were more than adequate compensations. Once again the electronic engineer was freed from detailed considerations and allowed to concentrate on the wider aspects of system design.

From the mid-1960s to the present day, this process of circuit accretion has continued and is likely to do so for the foreseeable future. The term 'silicon chip' became commonplace and what was available on two chips at any time would be available on one chip a year later.

Within this context the publicity accorded to the achievement of the single-chip microprocessor can be seen as somewhat misplaced. The microprocessor was but a natural step in this continuing process of miniaturisation. It has already been superseded in some applications by single chips bearing two microprocessors and ancillary circuitry besides.

Nor are the social implications of the microprocessor as severe a shock as the media would have us believe. These promises, or threats depending on your view, have been there since the first transistors were introduced and the more hysterical predictions of the dawning age of electronics echo those of the early 1950s.

This furore of publicity has tended to ignore or obscure the many other developments necessary before a microprocessor becomes a useable entity. Even ignoring the enormous problem of software, there is a whole range of auxiliary circuit chips necessary to support a microprocessor in any useful system architecture. The development of these, which is often more demanding than that of the microprocessor itself, has been pursued almost in the background and they are now available in bewildering variety.

As a consequence of this whole development process, the education of an electronic engineer today is required to be vastly different from that of, say, twenty years ago. Detailed circuit design is now the province of solid-state physicists as much as anyone; the electronic engineer concentrates on systems and system architectures.

The papers contained in this book describe the current trends in microelectronic engineering education. The syllabus of the subject is totally fluid and expanding daily, leading to obvious problems in providing industry with graduates who are familiar with modern technology and techniques. The papers contained represent the latest endeavours to this end and form what is effectively a status report of a continual and expanding activity.

A. A. BESWICK, Chief Engineer — Design Department, Ferranti Computer Systems Ltd, Manchester, England

Part 1

DESIGN AND FABRICATION OF INTEGRATED CIRCUITS

1

LOW-COST SILICON DEVICE FABRICATION IN DEGREE COURSES

M. J. MORANT
Department of Applied Physics and Electronics, University of Durham, England

1 INTRODUCTION

The majority of electronic engineering degree courses nowadays contain a considerable amount of semiconductor device and integrated circuit theory. This normally includes the basic operation of p-n junctions and transistors, manufacturing techniques such as oxidation, photolithography, and diffusion for discrete and integrated devices, and some simple IC design. This subject is one of the fundamentals of modern electronics and its study is an essential part of the long-term education of electronic engineers. Even though most graduates will be concerned with the systems applications of IC's, some background of the internal operation of chips gives at least an appreciation of the achievements and limitations of present-day devices and of future possibilities. Looking ahead, it is predicted that the advent of VLSI will lead to an increasing shortgage of chip designers so that the importance of silicon microelectronics in degree courses is unlikely to diminish.

To the student anxious to learn how to design useful systems, lectures on device theory may appear to be rather remote or even irrelevant. It is not easy to increase motivation by the addition of normal laboratory work because of the difficulty of devising really worthwhile experiments on IC's at chip level. Devices courses are often illustrated by physics experiments on semiconductor resistivity, Hall effect, drift mobility, etc. but these are too basic to stimulate much interest in microelectronics. A second approach is for students to examine opened IC's with a microscope and this can be far more productive particularly if the circuits are sufficiently simple to enable their operation to be related to the measured component dimensions. The early Type 702 operational amplifier was ideal for this since electrical measurements could also be made on isolated components using external connections and one or two probes put on to the chip itself. However, IC's as simple as the 702 scarcely exist nowadays so that this approach is of rather limited value.

A third approach to laboratory work on chip microelectronics is for students to make some simple silicon test samples themselves. A few university departments, fortunate to have silicon processing facilities for research, have used them for teaching in this way for many years. Others may be deterred by the

---

Based on a paper presented at the *Conference on Electronic Engineering in Degree Courses — Teaching for the 80's*, University of Hull, England, in March 1980.

high cost of current production equipment. However, for degree-level teaching with limited objectives, the equipment requirements are not insuperable, if there is plenty of workshop effort. This paper describes the facilities needed for the fabrication of simple MOS IC test chips that have been made by third year project students for the last three years. Suggestions are also made for the fabrication of arrays of simpler devices, such as MESA diodes, for which very little special equipment is required.

One objection to this type of degree-level laboratory work, is that it is too technological for academic courses. However, the subject of microelectronics has been driven by its manufacturing technology for the past twenty years and the design and functioning of chips is intimately related to processing methods so that the technology cannot and should not be avoided in teaching.

2 REQUIREMENTS OF SILICON PROCESSING FOR TEACHING

The growth of microelectronics has been brought about by reducing component dimensions and increasing the silicon throughput. Both have required increasingly sophisticated equipment which has been justified by improved production economics. In the teaching situation a great deal can be learnt by making a very small number of devices which need not have high performance, reliability, or yield. The main criterion of success is then the production of a few chips containing one or more components whose characteristics can be measured and interpreted in terms of the dimensions, material properties and processing parameters. It is certainly not necessary to process large pieces of silicon. A 4×6 array of devices or chips can be fitted easily on to a 10×20 mm silicon slice with sufficient success at 10% yield. This is such a reversal of the commercial situation that some of the equipment required becomes sufficiently simple for it to be made in a departmental workshop.

In the University of Durham we started making our own furnaces etc. for simple silicon planar processing, together with a home-made clean room in 1967. These basic facilities have been improved many times since then, more under the impetus of research than teaching. Over the years we made almost every item of equipment required for simple silicon technology and measurements, although much of it was later replaced by renovated second-hand industrial equipment. Elsewhere, entire teaching laboratories have been built up using secondhand industrial equipment, although at greater cost than the do-it-yourself approach[1]. We are now just entering the third stage of development with the completion of a commercial 1200 sq. ft. clean room suite and further refinement of existing equipment. This will be used equally for teaching and research.

The experience gained from the above has enabled us to reconsider what could have been done with just slight improvements to the original equipment, and this is the basis of the following description. However, it must be emphasised that the present results have actually been obtained using more secondhand than home-made equipment.

3 AN MOS TEST CHIP

For teaching, we have chosen to make MOS chips because they are simpler than bipolar structures and it is easier to relate device characteristics to theory. The simplest MOS process is p-channel, metal-gate, which requires only four masking stages to produce a range of self-isolated devices. However, there are three steps in MOS processing for which simple equipment may only just be adequate:

(a) the production of masks to resolve the channel length of no more than 25 μm to obtain convincing MOST characteristics,
(b) the growth of a sufficiently pure gate oxide to give a moderately low threshold voltage, and
(c) the alignment of the gate oxide with the source and drain diffusions.

None of these problems arise in bipolar device processing, but then epitaxial silicon slices are required and these are not so readily available. Epilayer growth cannot really be done safely at low cost.

An MOS teaching chip design is shown in Fig. 1 with a schematic in Fig. 2. The chip is designed to give a wide range of possible device measurements in the same way as one described by Beynon and Bloodworth[2] for similar applications. It contains seven large MOST's of three different channel width/length ratios, a capacitor, a diffused resistor, a Schottky diode, and a controlled

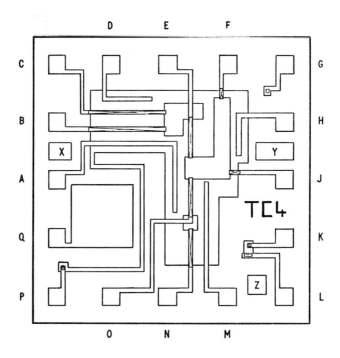

FIG. 1   Physical layout of MOS test chip TC 4. Actual size 1.80 mm square. X, Y and Z are alignment marks.

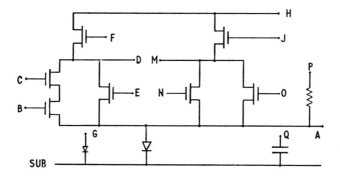

FIG. 2 *Circuit of test chip TC 4. Letters refer to the bonding pads in Fig. 1.*

avalanche diode. The alignment marks X, Y and Z are also p-n junctions. The chip is 1.80 mm square and the MOS channel length is 15 μm.

The chips are produced on $1\frac{1}{2}''$ diameter silicon slices, although far smaller rectangular pieces would be adequate. Because the manual step-and-repeat process for making the masks is very tedious, the number of chips is usually limited to a 5×5 array in the centre of the slice. This has provided more than sufficient devices for measurements.

4 PROCESSING EQUIPMENT

Although many textbooks give an outline of silicon microelectronic processing (e.g. Refs [3] and [4]) there is really no substitute for experience of the numerous problems that arise in practice. Some apparent details turn out to be critical for the avoidance of contamination. For student work, where simplicity and low cost are vital, it is therefore a matter of knowing what one can get away with both in processing and equipment. Such information is now more widely available outside the semiconductor industry.

The exact specification of equipment used will depend greatly on what is already available and what can be made. To buy everything new could certainly cost many thousand pounds even at this level of processing. However, many laboratories may already have equipment such as vacuum systems and microscopes used for other purposes, and most other items can be made if there are good workshop facilities. The cost of parts might then be as low as £2,000 spread over the construction period. Some details of the minimum equipment requirements are given below.

Clean room

It is probably worthwhile doing all silicon work in a separate room rather than an open laboratory. The room should be as clean as possible but it is not necessary to go to full clean-room standards. However, quite a good clean room can be constructed at much less than commercial cost if an existing room, such as a photographic dark room with forced ventilation and an air lock entrance, can be adapted. The room must be completely sealed and slightly pressurised with the inlet air finally passing through a HEPA filter. Everything inside must be made as particle-free as possible. Our own home-made clean room was developed over a long period of time but now regularly has an air particle count of only a few thousand particles greater than 0.5 μm diameter per cubic foot.

Even in a clean room, it is probably necessary to do the photolithography in a laminar flow hood. A four-foot vertical flow hood may contain two small photoresist spinners and a simple mask alignment machine. The furnaces can be mounted behind the hood with the tubes passing through holes in the back so that the samples can be loaded in a clean atmosphere immediately after etching.

Mask making

Whether it is worth trying to make masks will depend on the availability of photographic equipment. We were surprised to find that our laboratory plate camera with a 150 mm Simmar lens was suitable for the first stage reduction ($25 \times$). The 8" originals are cut in Rubylith on an ordinary printed circuit scribing machine and photographed on 2" square Kodak High Resolution Plates. The second reduction ($10 \times$) is done on a very early commercial manual step-and-repeat camera which is little more than a modified microscope, but other arrangements could probably be devised. The success in making MOST's will depend critically on the masks and it is well worth checking the resolution with a test pattern before proceeding too far.

Furnaces

Four furnaces are required for the p-channel metal-gate process. They are used for wet and dry oxidation, boron diffusion and annealing. For minimum cost the silica furnace tubes can be as small as 1" bore. Suitable furnaces can be made using resistance wire wound on to Mullite tubes. The temperature profile is not really critical for heat treating single small slices since a thermocouple may be placed very close to the sample, but the winding should be graded to reduce the sharp temperature peak in the centre of the furnace. The furnace elements are mounted in thermally insulated boxes. Normal electronic temperature controllers are adequate. High purity cylinder gases (N_2 and O_2) are suitable for the furnaces but they must be dried and filtered. It is not essential to measure the flow rate.

The most convenient source of boron for the diffusion is a slice of boron nitride if one can be obtained. The simplest alternative is an Emulsitone solution which is spun on to the silicon in the same way as photoresist.

Photolithography

The mask alignment and exposure machine is probably the most crucial piece of apparatus because the gate oxide window must be accurately positioned between the source and drain diffusions. The photoresist coated silicon is mounted on a vacuum chuck capable of very fine X-Y movement. The mask should be mounted rigidly as close as possible above the silicon but without contact. A microscope is used for observing the silicon surface through the mask, enabling the patterns to be lined up. A laboratory UV lamp can be used for exposure with the light directed through the mask from a surface-aluminised mirror which can be swung into place below the microscope once the alignment is satisfactory. There is no need for the refinements found in commercial alignment machines because of the small number of chips on the silicon and because there is plenty of time for manual adjustment. However, the X-Y slides must be capable of smooth movement of a few μm.

Metallisation
Good quality high vacuum evaporation systems are needed for the deposition of the aluminium metallisation on the top of the slice and of gold-antimony for the ohmic back contact. Liquid nitrogen traps are probably essential to obtain a sufficiently clean vacuum.

5 MEASUREMENTS

Diagnostic measurements
To compare device characteristics with theory it is necessary to measure some of the physical parameters of the processing. These 'diagnostic measurements' also serve to monitor each stage. For such measurements a fresh silicon test slice is included with the device slice at each stage of processing. The following physical measurements are possible using largely home-made equipment:
(a) Resistivity of initial silicon from a 4-probe resistance measurement.
(b) Sheet resistance of p-type layer from 4-probe measurements.
(c) Thermoelectric typing of both initial and diffused surfaces.
(d) Junction depth measurement by grooving and staining.
(e) Thickness of masking and gate oxides from the zero-bias capacitance of a dot of aluminium evaporated on to the oxide through a metal mask. The thickness can also be estimated from the colour of the films.
(f) Mercury probe C-V technique to measure the diffusion profile of the boron.
(g) Quality of the gate oxide assessed from the stability and hysteresis of the small-signal C-V characteristic of the capacitor on the chip. In the absence of a C-V plotter a normal r.f. bridge with a simple circuit for applying a dc bias is adequate. The results will probably show the presence of mobile positive ions in the oxide which is quite unacceptable in commercial devices but not disastrous in the teaching context.

Device measurements
The device measurements are made on the chip by putting probes on to the appropriate bonding pads. The probes should be rigidly mounted on micromanipulators attached to a microscope stage. The following electrical measurements can be made on the devices:

(a) MOST *I-V* characteristics. By limiting the negative dc gate voltage, the threshold voltage can be kept sufficiently stable during the measurement process even if the transistors would be useless for any real applications.
(b) Gate and bistable operation. The chip of Fig. 1 has transistors connected as NAND and NOR gates. With external connections the gates can be cross-coupled to form a bistable. Naturally, the operating speed will be extremely low because of the large capacitances.
(c) Schottky barrier and p-n junction *I-V* and reverse bias *C-V* characteristics.
(d) Diffused resistor *I-V* characteristics.

The results obtained in three years of making chips similar to Fig. 1 have been very encouraging. Many of the MOST's have electrical characteristics close to the theoretical expectation based on the dimensions and diagnostic measurements. However, we have not so far had a sufficiently high yield to check the gate operation. The diodes, particularly the p-n junction type, have more reliable characteristics, and if all else fails on a chip, the diffused resistor can be relied upon to give the expected value of about 400 ohms. It is not worth quoting other results because they are so dependent on processing care. For research, far better results can be obtained as the processing time is effectively unlimited.

6 ALTERNATIVE DEVICE STRUCTURES

Previous sections have described the fabrications of the most elaborate chip likely to be required for undergraduate teaching. Simpler device structures can, however, be made with less equipment. The following suggestions are in order of increasing complexity and they might be used as intermediate stages to the eventual implementation of an MOS teaching facility.

Schottky diode arrays
Rather poor Schottky diodes can be made on n-type silicon by evaporating aluminium dots through a contact mask and annealing at about 490°C. A large area gold back contact is also deposited by evaporation. Very little special equipment is required but the diodes have far from ideal *I-V* characteristics, although the reverse bias *C-V* agrees well with theory.

Diffused Mesa diode arrays
Mesa diodes are produced by diffusion of boron over the entire surface of an n-type wafer followed by a masking process and etching of the silicon to isolate the individual junctions. The masking can be done by depositing drops of suitable varnish on the surface, although photoresist is obviously better. The p-n junctions should have good electrical characteristics and many of the diagnostic measurements listed in Section 5 can be made on test slices. This is therefore a good level to attempt because the equipment required is far less than for MOS and even a clean room is not essential.

Diffused planar diode and resistor arrays
The planar process for making p-n junctions is based on oxide masking and photoresist. A wet oxidation furnace, photoresist spinner and laminar flow cabinet are therefore needed. However, there is little to be gained over the previous level in measurements except for those on diffused resistors. The main benefit to students is in the use of oxide masking which is one of the most basic processes in micro-electronic technology.

7 STUDENT REACTION
Our only experience of students making silicon devices has been with MOS chips similar to Fig. 1. These have been made by pairs of students in Durham as one-term third year projects for the last three years. The mask making and some of the processing has been done by a technician, because of the time required, but with sufficient student involvement for them to be aware of all the problems. Each year far more good MOST's and other devices have been produced than could be measured thoroughly. The students have been notably enthusiastic about these projects and have benefitted from learning of some of the problems and limitations inherent in all IC production. Although a particularly valuable experience for those thinking of a career in microelectronics, it also broadens the outlook of the more applications-oriented student and his appreciation of the very real achievements of LSI.

It is certainly not necessary for every student in a class to make silicon devices since the benefits rub off from one to another. We will need to experiment further with varying degrees of student involvement in processing, perhaps making it an option for those who feel that they might go into the semiconductor industry, while a larger proportion do some of the measurements.

8 CONCLUSIONS
It is hoped that this paper may encourage others to attempt some silicon device fabrication in undergraduate courses. If considerable time and effort can be devoted to building and developing simple teaching-level equipment, very large amounts of capital become unnecessary. However, the running costs should not be overlooked since very pure consumable materials have to be used. The paper has outlined the probable minimum requirements for an electronics teaching facility and it may horrify those concerned with more professional processing. Although this level of technology is effective in undergraduate teaching it is insufficient for more advanced courses. The facilities at Durham are therefore being improved to increase the resolution, yield and reliability for post-graduate teaching and research.

It is concluded that the benefits of some silicon device fabrication in undergraduate courses need not be restricted to the larger departments if time, effort, and a comparatively small amount of money can be allocated to building up a simple microelectronics processing laboratory.

REFERENCES

[1] Anner, G. E. 'Laboratory for Fabrication of Solid State Devices' *1978 Frontiers in Education Conference Proceedings, IEEE, New York*, pp. 133–135 (1978).
[2] Beynon, J. D. E. and Bloodworth, G. G. 'A Practical Approach to the Teaching of Integrated Circuit Design' *Proceedings of the Conference on the Teaching of Electronic Engineering in Degree Courses, University of Hull* Paper 9 (1973).
[3] Burger, R. A. and Donovan, R. P. *Fundamentals of Silicon Integrated Devices Vol. 1*, Prentice-Hall (1967).
[4] Glaser, A. B. and Subak-Sharpe, G. E., *Integrated Circuit Engineering* Addison-Wesley (1977).

2

A NOVEL UNIVERSITY APPROACH TO TEACHING MICROELECTRONICS

K. F. POOLE
Electronic Engineering Department, University of Natal, Durban, South Africa

1 INTRODUCTION

Over the past five years, the nature of the activity in the field of microelectronics in South Africa has changed from one of essentially a user to that of a producer. This activity has accelerated at a high rate and over a wide sphere, with industry, research institutes and universities becoming increasingly involved in the design, manufacture, technology and applications of microelectronic circuits. To supply the manpower to support these activities the universities embarked on programmes geared to train engineers with a capability in the field of microelectronics.

A major development, underway in 1975, was the production of an uncommitted integrated circuit[1], the UC-1, by the local Council for Scientific and Industrial Research (C.S.I.R.) in their Electrical Engineering Institute. This highlighted the lack of design capability in the country.

The approach taken over the past four years to train integrated circuit design engineers and thus meet the major demands was based on the following requirements and philosophy.

(i) The support of locally-manufactured products and the promotion of these products into industry. It was felt that young graduate engineers with confidence in handling the uncommitted chip is one of the most successful means of launching this concept and product into industry. Courses leading to this end are consistent with the policy to have a strong link with the manpower requirements of industry and to foster new technologies and modern electronics in industry.

(ii) Co-operation at all levels between organisations involved in this expensive technology is sought after and by actively supporting a local product there has been good co-operation with the C.S.I.R. and the personnel of the Electrical Engineering Institute on this programme.

(iii) A fundamental policy towards the teaching of microelectronics was adopted and hence a careful balance between the technological aspect and the theoretical course material was aimed at.

2 THE DEGREE STRUCTURE

The duration of the B.Sc. (Eng) degree course is four years with each year divided into two semesters. The full curriculae are given in Appendix 1 together with the approximate number of hours associated with a full-credit and a half-credit course.

All Electronic Engineering students take a half-credit course in Electrical Materials in their second year of study. The purpose of this course is to cover the basic electrical properties of materials and, in particular, fundamental semiconductor theory.

During their third year of study, students attend a one-credit course in Physical Electronics which is biased towards solid state devices and integrated circuit theory.

In final year the half-credit course in Physical Electronics is one of a number of electives available, and one of the topics dealt with in detail in this course is the uncommitted chip. The one and a half-credit design project may also be used to study microelectronics, provided the student has completed the half-credit elective course described first.

In conjunction with these courses, Electronic Circuit Design is taught in the electronics courses offered in the third and final year.

3 THE UNCOMMITTED CHIP

Essentially, an uncommitted chip consists of a large number of electronic components, manufactured in an integrated technology, with all the contact pads short-circuited by means of a continuous metal layer covering the entire chip. By the selective removal of metal, specific components are left interconnected and various electronic circuit functions can be realised.

A standard bipolar technology is used in the manufacture of the UC-1. It is 3 mm × 3 mm in size and contains over three hundred components including transistors, npn and pnp, diodes, resistors and capacitors. A complete listing is contained in Appendix 2.

The process utilises 50 mm dia. wafers and hence each wafer contains approximately 200 chips. The wafers are coated with aluminium as the conductor material and are supplied by the C.S.I.R. in this form.

With information[2] on device characteristics, tolerances etc. available in a designer's handbook, the design engineer's task is to design a circuit to fulfil the desired function, making use of the components on the chip. The design completed, it is then necessary to interconnect the required components according to the circuit diagram.

This is achieved by using the layout sheet, supplied by the C.S.I.R. This 750 mm by 750 mm sheet is an accurate and stable reproduction of all the component connection pads on the chip and, using 3 mm tape, for example printed-circuit-board tape, it is possible to connect up the desired components as per the circuit diagram. Apart from obeying certain layout rules, this is a comparatively simple step.

This large scale circuit is then photographically reduced to the correct size by means of a first reduction camera (approximately 25 × reduction) and a step and repeat camera which not only reduces the size by approximately 10 × but also produces an accurate x-y array of identical images. The step distances along the x and y axes can be accurately controlled, as can the number of rows and columns. This is set to match the array of chips on the silicon slice.

The uncommitted wafer is then covered with photo resist, the resist exposed through the photographic mask after alignment on the mask aligner and then the resist is developed. The resist which remains after developing and post baking forms a hardened protective layer over the aluminium and prevents etching. Finally, the unwanted aluminium is etched off and only the aluminium strips interconnecting the required components are left. After stripping the rest of the photo resist, the wafer is sintered.

Finally, the wafers are scribed, broken into chips, mounted in packages and bonded out ready for testing.

4 THE MICROELECTRONICS FACILITY

The broad objectives are typical of most universities[3], that is, to primarily train undergraduate students, to cater for postgraduate research students and to act as a local facility for the industrial organisations located within a 100 km radius of the university. In this regard it should be mentioned that the university is situated 700 km from the C.S.I.R. and it is thus favourably placed to offer a service to companies in its neighbourhood.

Coupled with this initial programme was also the future developments which could take place in order to increase the capabilities of this facility to cater for the more fundamental and basic research into devices as and when it was required.

Complementing the uncommitted chip programme is the work on thin film hybrid electronics and the laboratory equipment is used for projects in this direction as well.

4.1 *Post graduate research and industry*

As this is a new facility which is still under development, postgraduate work is in its infancy and in fact this year saw the first M.Sc. student registered for work in this field. The 1979 graduates were also the first engineers with uncommitted chip experience to be employed in industry and it is heartening to see that already arrangements have been concluded to have ex-students back for short periods of time to make use of the facility and test out potential circuits for industrial applications.

Thus the original concept and objectives of the facility have now been realised and it is rewarding to see such a positive and fast response by industry.

The advantages to industry, over and above the advantages of size, reliability etc, attendant with the introduction of integrated circuits are those of short turn-around time for development work, a nucleus of expertise to draw on and finally the overall cost savings.

4.2 *Undergraduate teaching*

As an undergraduate teaching facility the objectives were that students could:-
(i) design an integrated circuit and have a facility to test their design; the

testing was deemed to be most important.
(ii) the turn-around time between the completion of a design and the final product in a form suitable for testing had to be less than seven days and of the order of 40 man hours.
(iii) the cost of this exercise should be low, less than R50 (U.S. $40) in order that a number of students could make use of the facility and test more than one design, should the first attempt prove unsatisfactory.

The semester during which the Physical Electronics course, and hence this activity, runs is nine weeks long and the final design project extends over a period of four weeks, hence the requirement for a fast turn-around time.

In final year the course in Physical Electronics has associated with it one practical project and all students taking the course are involved in using the apparatus in the microelectronics laboratory to familiarise themselves with the equipment. This necessarily entails close supervision. Due to the numbers involved, students usually work in groups of two or three.

The university microelectronics laboratory is equipped with the essential apparatus and services to perform the final stages necessary for turning an uncommitted integrated circuit into a useful operational integrated circuit.

It is essential in such a laboratory to have small numbers of students if hands-on experience is to be gained by the students, and the author feels strongly that this situation should be maintained.

Students work individually on a specific project during the final four weeks of their degree course. The area of interest selected by a student is generally accommodated, but the specific details are drawn up by the responsible staff member. The facility is limited to four students during this time and the projects undertaken in 1979 were as follows:

(i) *UC-1 digital gates — I. Clark.* The emphasis in this project was directed at the process. Over the previous semesters a set of somewhat arbitrary procedures had evolved. This project set out to optimise the process steps, paying particular attention to any critical steps, and then a set of working rules, designed to make the task of future students a great deal easier, was compiled. Combined with this was the requirement to investigate the use of the UC-1 for digital applications.

(ii) *UC-1 Phase locked loop — D. Knee.* The aim of this project was to manufacture a phase locked loop integrated circuit, using the UC-1, and having specifications similar to the commercially available type 565 chip. Detailed design calculations and final testing was required.

(iii) *Custom integrated circuits versus discrete components — N. Hansen.* The student was presented with a working discrete component module which included an attenuator and audio amplifier. The project required the circuit to be redesigned and built as an integrated circuit using the UC-1. To conclude the project a set of test results comparing the performance of both circuits was required.

(iv) *Thin film D/A converter — G. C. Hesse.* The aim of this project was to design and manufacture a thin film resistor array suitable for a 4 bit D to A converter and test a prototype.

4.3 Apparatus and facilities

The equipping of the laboratory followed a fairly standard pattern in the university, with equipment being obtained in a variety of ways, including straight purchases using grants, university capital etc., apparatus on loan and donations.

Already in operation was a laboratory for the preparation of silicon slices and hence the associated diamond saw, ultra pure water supply, hand lapping and polishing facilities were available together with chemical handling facilities such as fume cupboards etc. A vacuum system for coating the silicon wafer with aluminium, stereo microscope and other laboratory apparatus was also on hand.

Thus the major items of equipment required were:

(i) First reduction camera. This was built in the department using a NORITAR f3.5, 70 mm lens, and a light box employing back neon tube lighting through a green filter, to suit the film sensitivity, was constructed. Vacuum is used to flatten the masks onto the glass front of the light box.

(ii) Spinner and scriber. These were both manufactured by the university. A standard diamond scribe tool is used in the manual scriber.

(iii) A mask aligner and exposure station was loaned to the university by the courtesy of S.T.C., a member of the Altech group of companies.

(iv) The step and repeat camera was donated to the university, as was a rubylith cutting table for use in the thin film work, by the C.S.I.R.

(v) The bonder and prober were bought by the university, as were the fans and Hepa filters required to manufacture clean air cupboards and hoods.

(vi) A metallurgical microscope with camera attachment is on the budget for this year.

A standard photographic dark room was available. The only addition made here was filtered water. Also, a valuable resource on the campus for this work is the scanning electron microscope.

The facility requires the services of a technician in order to run efficiently. To date there is no staff for this purpose.

5 PROCEDURE

The uncommitted chip is supplied as a 50 mm diameter wafer by the C.S.I.R. The first step is to scribe and break this into four equal quarters, giving the student a four by five array of good chips or a possible 20 circuits. Taking into account a low yield, (usually better than 30%) this is more than adequate for test purposes. Fig. 1 shows a photograph of the wafer or slice as used by students. The unused corners are useful for handling the slices which are stored in a vacuum desiccator.

To reduce the cost of the photography, a high quality polyester-based film was experimented with and has been found suitable. Compared to glass slides which are difficult to obtain in small quantities and require involved processing, the polyester film requires no extraordinary processing. A simple developing and fixing bath is required and the cost of the film and chemicals is well

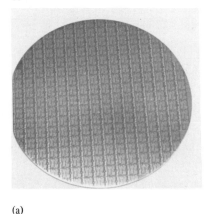 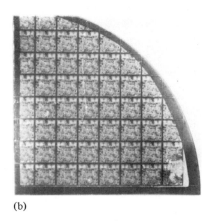

(a) (b)

FIG. 1   (a) 50 mm diameter wafer (b) processed slice.

below that for glass slides. Thus, at the mask-making stage, students are encouraged to take great care and repeat this stage many times to ensure that no errors are present as this is an inexpensive procedure. Careful optical checking is carried out for gross errors such as missing connections, dust or scratches. With this simple photographic process, little trouble is experienced by the student and it can be completed in two hours.

After the first reduction, the negative is used to make a contact print which forms the input to the step and repeat camera. A delay of some three hours is necessary here to ensure that the negative is properly dry. Typical reproductions are shown in Figs. 2 and 3.

The step and repeat camera produces the final six-by-six array and, once again, careful checking is performed to ensure that this negative is of good

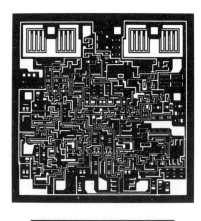 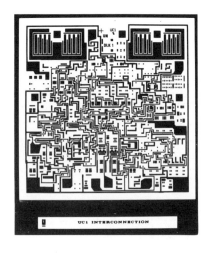

FIG. 2   First reduction (approx. full size).

FIG. 3   Contact print (approx. full size).

quality. This step takes approximately two hours, and the negative is left to dry overnight. A print of the array is shown in Fig. 4.

Up to this stage, all the above steps can be completed comfortably in one day.

The slice is then removed from the storage desiccator and placed on the spinner. Dust is a major problem and although most of the work is carried out under clean-air hoods and in clean-air cabinets, before any critical stage, the surfaces are blown with 'dust free' to make sure no dust particles are left to cause a problem.

The slice is coated with photo resist using a filtered manual dispenser and spun for 30 seconds at 5000 r.p.m. The prebake is carried out in an oven at 70–80°C for 20 minutes. Care is also taken to ensure that the effect of ultra violet rays from stray light is at a minimum during these process steps. The room is illuminated with yellow light and as the mask aligner is some distance from the spinner, the coated slices are carried in light-tight boxes.

The step and repeat array negative is inserted into the mask aligner and blown with 'dust free'. The slice is placed on the mask align chuck, the two are aligned, clamped and exposed. This process usually takes about 15 minutes.

Another day is required to complete the process up to this point.

At this stage, an optical inspection of the slice is performed in order to eliminate slices which may have suffered mechanical damage, alignment errors or any other catastrophe which renders the complete slice reject. This step is particularly important, as one of the innovations introduced in the university facility is the recycling of these slices thereby eliminating waste and reducing costs.

If the process is stopped before sintering, the aluminium can be completely removed, the silicon slice re-etched and finally a new layer of aluminium

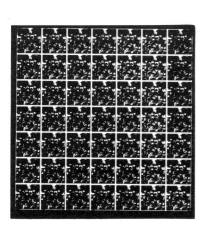

FIG.4 Final array (approx. 2 × full size).

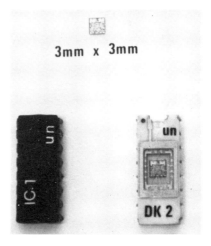

FIG.5 Chip and 16 pin dual-in-line packages (approx. full size).

deposited over the complete slice. The slice is then ready to pass through the process again.

This cycling has been successfully performed up to three times on one slice thus giving the student four attempts on one slice up to this stage in the process! This potential saving encourages careful inspection at the optical stage before sintering. This inspection is laborious and can take 30 minutes or more depending on the circuit complexity.

If the yield optically is satisfactory, the wafer is sintered for 15 minutes in dry nitrogen at 475°C. After sintering, the circuits are probed and simple electrical tests used as a means of selecting potentially good working circuits.

Probing is also a fairly time-consuming activity but can yield useful initial test data and does eliminate packaging non-operational circuits. The time taken depends on the complexity of the circuit and the number probed. The slice is then scribed and broken into chips. The selected chips are then stored ready for packaging. All chips carry their own code identification and each wafer is also coded with a batch number. This information is retained with each chip in order to maintain a record of the history.

A photograph of a chip and packaging is shown in Fig. 5.

Another day is required to reach this stage and one final day is required for the chip bonding, wire bonding and final packaging.

Although ceramic packages are expensive, they allow easy removal of the cover plate should it be required to check the chip at a later stage for failure mode analysis or any other reason. This has some advantages for teaching purposes.

A check on quality assurance is maintained by sending a sample of each slice to the scanning electron microscope unit for more detailed photographs. These reveal alignment and size errors as well as process defects such as etching and photo resist problems.

The task of layout and taping the master circuit is time-consuming and adds on an extra two days to the total of four so far accumulated. This time schedule is thus within the requirement of section 4.2.

An estimate of the cost of the materials used by a student to process one circuit is not straightforward. However it is certainly well within the R50 (U.S. $40) limit aimed at.

6 CONCLUSION

This laboratory has provided the university with a low-cost facility capable of training electronic engineers in the design of integrated circuits at the circuit design level. As the future requirements for manpower at a more fundamental level of chip design grow, it will be a relatively easy task to add to the facility the necessary diffusion ovens etc., and hence increase the capability of the facility to cater for this need as well.

The uncommitted chip provides an excellent vehicle for the training of undergraduate electronic engineers in the field of microelectronics. It relieves the student of some of the more exacting and often more hazardous technology whilst exposing them to sufficient technology to give them an awareness of what is required and to provide a stimulating course with a balance between the practical and theoretical material.

The author has not enquired of other semi-custom chip manufacturers whether they would be prepared to sell wafers in the form required. This should be no problem as this activity acts to increase the usage of this concept and advertises the uncommitted chip to many students who will be tomorrow's users in industry. It in no way removes part of or any prospective custom from the chip manufacturer, as the university facility is not designed to handle large quantities but rather is a one-off testing facility and, in this way, actually complements a factory. The modus operandi is that should industry be satisfied with a prototype developed by the university, then the C.S.I.R. would handle the bulk order with the advantage of knowing that the original layout has already been tested and therefore no errors of this type will arise.

Although this approach was specifically tailored to meet a particular need, the concept of using the uncommitted chip approach for teaching purposes is one which the author feels could be applied by any university engaged in the teaching of microelectronics.

7 ACKNOWLEDGEMENTS

With a facility this large, acknowledgements are due to many people and organisations and I would like to thank local firms and people who generously supplied goods at cost or even donated them in many instances.

A special thanks to the C.S.I.R. for their support, especially the members of the Solid State Group.

Thank you to Altech for the loan of the mask aligner without which we could not have continued, to Mr. Johnson of the university workshops who manufactured most of the items made by the university, to the University Research Fund for financial support and to all students from 1975 onwards who have added to the facility, in particular Mr. F. Nunneley who set up most of the equipment.

And finally the author acknowledges the inspiration and encouragement derived from the paper *An economical microelectronics laboratory* by Prof. T. K. Gaylord, see Ref. [3].

8 REFERENCES

[1] *The design of custom integrated circuits for local manufacture;* a Design Course held on the 20/21 October 1975 at the C.S.I.R. with lectures by members of the Solid State Division of NEERI.

[2] Mathlemer, W., *The uncommitted integrated circuit designer's handbook.* The Solid State Electronics Division, NEERI, C.S.I.R., P.O. Box 395, Pretoria, S. Africa.

[3] Gaylord, T. K., 'An economical microelectronics laboratory'. *Microelectronics* **5**, No. 1 (1973).

9 ADDENDUM

Recent developments in microelectronics activities

The concept of using uncommitted chips for breadboarding in silicon has worked well and uncommitted chips provide the vehicle on which a large programme for teaching I.C. design is based.

The following points briefly summarise the developments which have occurred since January 1981, when this paper was first published.

(i) The post graduate school now has seven M.Sc. students.
(ii) The use of analysis and simulation programmes is taught to undergraduates whilst much of the circuit development at post graduate level relies on this approach.
(iii) Breadboarding large circuits (500–1000 components) using a number of uncommitted chips containing 100–200 components combined with the theoretical analysis forms part of the design cycle which has evolved.
(iv) The use of CAD layout, DRC packages and photo-plotter has reduced the mask-making effort.
(v) A number of uncommitted chips have been designed and are now in use. The UC-3 and UC-4 have more components and are also available in a higher voltage process.
(vi) MOS gate array chips, currently being designed by post graduates will be available shortly. A 200–300 NMOS gate array and a 200–300 CMOS cell array will be available as standard uncommitted chips.
(vii) Final year courses in Microelectronics are now given in both semesters.

APPENDIX 1

Course of study required for B.Sc. (Eng)

First year

First semester	Credit value	Second semester	Credit value
Calculus	1	Calculus	1
Vectors and Matrices	1	Mechanics	1
Chemistry	1	Basic Organic and Additional Physical Chemistry	$\frac{1}{2}$
Physics	1	Topics in Inorganic Chemistry for Engineers	$\frac{1}{2}$
Engineering Drawing	$\frac{1}{2}$	Physics	1
		Eng. Design (Electrical)	$\frac{1}{2}$

Second year

First semester		Second semester	
Mathematics II(a) Eng.	1	Mathematics II(b) Eng.	1
Theoretical Mechanics	$\frac{1}{2}$	Electricity	1
Computer Methods I	$\frac{1}{2}$	Engineering Design (Elec)	1
Electricity	1	Electrical Materials	$\frac{1}{2}$
Strength of Materials	$\frac{1}{2}$	Computer Methods II	$\frac{1}{2}$
Optics and Wave Motion	$\frac{1}{2}$	Atomic Physics	$\frac{1}{2}$
Thermo-Fluids	$\frac{1}{2}$		

Third year

First semester		Second semester	
Part. Diff. Equations	1	Numerical Methods	$\frac{1}{2}$
Elec. Design (Heavy)	$\frac{1}{2}$	Circuit Theory	$\frac{1}{2}$
Circuit Theory	$\frac{1}{2}$	Control Systems	$\frac{1}{2}$
Electrical Machines	1	Electronics	$\frac{1}{2}$
Elec. Measurements	$\frac{1}{2}$	Elec. Design (Light)	$\frac{1}{2}$
Electronics	$\frac{1}{2}$	Physical Electronics	1
		Plus *two* from the following options:	
		E.M. Field Theory	$\frac{1}{2}$
		Information Systems	$\frac{1}{2}$
		Power Systems	$\frac{1}{2}$

Note: Information Systems is a prerequisite for Electronic Engineering. E.M. Field Theory is a prerequisite for Communications, Microwave Field Theory and Microwave Engineering.

Fourth year

First semester		Second Semester	
Electronics	$\frac{1}{2}$	Electronics	$\frac{1}{2}$
Electrical Design (L)	1	Design Project	$1\frac{1}{2}$
Macro Economics	$\frac{1}{2}$	Professional Practice	$\frac{1}{2}$
Plus *four* from the following options:		Plus *three* from the following options:	
One free choice of $\frac{1}{2}$ credit course	$\frac{1}{2}$	One free choice of $\frac{1}{2}$ credit course	$\frac{1}{2}$
Filter Networks	$\frac{1}{2}$	Physical Electronics	$\frac{1}{2}$

continued

Control Systems	$\frac{1}{2}$	Control Systems	$\frac{1}{2}$
Illumination Engineering	$\frac{1}{2}$	Illumination Optics	$\frac{1}{2}$
Elec. Communications	$\frac{1}{2}$	Elec. Communications	$\frac{1}{2}$
Digital Processes	$\frac{1}{2}$	Digital Processes	$\frac{1}{2}$
Microwave Field Theory	$\frac{1}{2}$	Microwave Engineering	$\frac{1}{2}$
Acoustics	$\frac{1}{2}$	Power Electronics	$\frac{1}{2}$

The hours per semester per course vary depending on the course and hence the figures given below are representative only.

	Hours 1 credit	Hours $\frac{1}{2}$ credit
Lectures	30	20
Laboratories	20	20
Tutorials	10	5
Assignments	20	20

APPENDIX 2

Description of Components Available in UC-1

List of UC-1 components

Description	Number present per chip
Small NPN transistors	63
Medium NPN transistors	16
MB36 NPN transistor	1
Test NPN transistor	1
Power NPN transistors	2
Total NPN transistors	83
Lateral PNP transistors	19
Double lateral PNP transistors	5
Substrate PNP transistors	4
Field-aided PNP transistors	2
Total PNP transistors	35
Schottky diodes	8
Capacitors (30 pF)	3
I^2L array	1
Hall-effect element	1
Pinch resistors (150K + 35K)	6

Total pinch resistance: 1,1 M ohm

Diffused resistors:	3K6 ohm	32
	1K8 ohm	38
	900 ohm	44
	450 ohm	45
	200 ohm	25

Total resistance:	248 k ohm	
Total number of components		327
Bonding pads		16 fixed
		9 optional
− V connections		15
+ V connections		13

3

A CMOS BREADBOARD GATE ARRAY FOR EDUCATIONAL AND R. AND D. PURPOSES

S. L. HURST
School of Electrical Engineering, University of Bath, England

1 INTRODUCTION

In the area of higher education there is increasing emphasis upon the desirability to teach both the principles and practice of microelectronic circuit design. The practical side should, ideally, involve the fabrication of student's designs, implemented both cheaply and rapidly, which, for most educational establishments, means the availability of some multi-project* or standard gate-array chip approach, fabricated by outside resources.

The complexity of design required at this educational level is not normally high. Hundreds rather than thousands of gates should suffice for student designs, since time will be a major controlling factor. Equally, such a size product also caters for many other research and development (R. and D.) project requirements which occur in university and other research activities. Hence the availability of a means of fabrication of relatively simple microelectronic circuits, not involving the limits of existing technology or requiring any time-consuming hand-crafted layout, is of wide significance.

The custom-design of a microelectronic circuit may follow one of the following three main routes shown in Fig. 1.

If we briefly compare the pro's and con's of the full custom and semi-custom design paths, we find the following characteristics.

(a) *Full custom design:*
   Most compact and efficient final circuit; however, very expensive and lengthy design procedure. Not justified for small quantities except for very special circumstances such as military or avionics;

(b) *Semi-custom design:*
   Cheap and efficient methods of producing special L.S.I. circuits; comparable to the design of a printed circuit board of a decade ago, but now fabricated on a single I.C. chip.

Clearly for educational and similar area applications it is route (b) rather than (a) which is relevant.

---

*A multi-project chip is a large-scale integrated (LSI) circuit upon which a number of separate project designs has been assembled and fabricated. Different bonding prior to final packaging is arranged such that only one of the several designs is connected per package, with selective delivery back to each project designer.

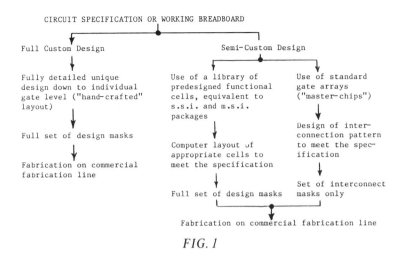

FIG. 1

When we consider further the two paths in the semi-custom design route shown above, then the adoption of the standard gate array approach has advantages in that only the interconnection design and fabrication has to be undertaken for any given custom requirement. This approach may also be referred to as the master-chip approach, since quantity fabrication of a master L.S.I. chip containing a large array of identical but not interconnected cells can be undertaken, the dedication of the cell array to realise any specific small-quantity design requirement being by the cell interconnection pattern design[1-5].

The master-chip approach has no educational disadvantages, since circuit design rather than silicon fabrication technology is the area of significance. Thus a standard uncommitted chip, requiring only the final interconnect design and fabrication for its commitment, is an ideal vehicle for educational and simple R. and D. applications.

The choice of standardised items provided on an uncommitted chip is clearly of importance if the chip is to have wide application. The uncommitted-logic-array (ULA) approach of several commercial products, wherein the uncommitted chip contains an array of standard cells, each cell being a small assemblage of separate components, has obvious attractions, since complete flexibility of interconnection down to individual transistor level is available. Against this advantage must be weighed the fact that all interconnections, both *within* each cell to interconnect the component parts, and *between* cells to form the complete system, are involved at the commitment stage.

Standardisation upon functional cells (gates) on the uncommitted chip, giving an uncommitted-gate-array (UGA) master chip, therefore has advantages from the amount of commitment design necessary, which is particularly significant in a time-limited educational environment. Indeed, the ULA and the UGA approach may be loosely likened to a project design built (i) using discrete components and (ii) using functional SSI (small-scale-integrated) gates, respectively. However, the choice of the functional entity on the uncommitted chip is clearly of supreme significance.

Gate arrays are available using NAND gates as the standard cell. Clearly all combinatorial and sequential digital functions are available from appropriate NAND gate assemblies, using standard textbook design methods. However, NAND and other conventional Boolean gates have very limited logical capability per gate, thus necessitating the use of numbers of such gates in simple applications. In contrast to such basic gates it is possible to formulate logic configurations which have a greater capability, and in particular configurations known as universal-logic-gates (ULGs) which are capable of realising all possible functions of a given number of input variables, depending upon the pattern of their input connections.

In this paper we will introduce the subject of universal-logic-gates, and show that a number of possible universal circuits can be proposed, each of which is characterised by being able to realise a complete range of logic functions without any internal circuit modification. They therefore represent ideal candidates to consider for adoption as the standard uncommitted gate array; the fear of initial unfamiliarity to a designer will be commented upon in Section 6.

We will therefore concentrate in the following sections upon the extreme right-hand possibility shown in Fig. 2, this diagram summarising the main cell structures that can be proposed for master-chip L.S.I. packages.

2 UNIVERSAL-LOGIC-GATES

Universal-logic-gates, sometimes termed universal-logic-modules (ULMs), are combinatorial circuits which are capable of realising all possible functions of a given number of input variables, depending upon the pattern of connections made to them[6-10]. A ULG circuit which is capable of realising all functions of two input variables A, B is termed a ULG.2 circuit; in general, a ULG circuit which is capable of realising all functions of n input variables is termed a ULG.n. Note that a ULG.n circuit is always capable of realising all functions of less than n input variables, and hence has $\leq n$ universality.

The operations which may be required in order for a ULG to realise a given combinatorial requirement are as follows:
(i) external negation (N) of any one or more of the binary input signals into the gate,
(ii) external permutation (P) of any two or more of the binary input signals into the gate,

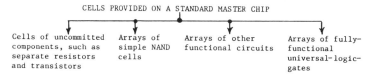

FIG. 2

and
(iii) external negation (N) of the output from the gate.

If, with a given ULG circuit, it is necessary to evoke all three operations (i), (ii) and (iii) in order to realise all possible functions of n binary input variables, then such a ULG circuit is termed a

NPN-complete ULG.n circuit

However, if external negation of the output is unnecessary, then such a ULG circuit is termed a

NP-complete ULG.n circuit

If output negation but no input negations are necessary, then we have a

PN-complete ULG.n circuit

Finally, if no external negations at inputs or output are required, we have a

P-complete ULG.n circuit

Thus, permutation of the input connections in order to generate dissimilar functions is the most significant operation with ULG circuits, and it is this feature which distinguishes them from our more familiar Boolean AND, OR, NAND and NOR gates which have complete functional symmetry in their input connections, and perform the same duty no matter how their inputs are connected.

The formulation of ULG circuits has been extensively investigated. The specification of P-complete or NP-complete ULG circuits with the minimum of input terminals per circuit has been the principal goal. It has been shown that for NP-complete ULG.2's, a minimum of three input terminals per circuit is necessary; for NP-complete ULG.3's, a minimum of five input terminals is necessary. These results represent the best achievements that are possible[11]. One additional terminal is always required to convert any NP-complete ULG into a P-complete ULG.

The range of possible ULG.2 circuits has been exhaustively investigated, particularly because this size of ULG rather than the larger ULG.3 may be preferable to adopt as a standard logic cell. It has been shown that there are only six possible basic ULG.2 circuit topologies with the minimum of three input terminals, these six candidates being illustrated in Fig. 3. Several circuit variants of each candidate, however, are available, since any input and the output of each circuit may be freely complemented from that shown in Fig. 3 without impairing the universality or classification of the circuit.

Further consideration of these six basic configurations shows that f_I, f_J and f_K are NPN-complete ULG.2's, but f_H, f_L and f_M are NP-complete[11]. Thus, the latter three are clearly more successful candidates than the former set of three. There remains the final choice of which one to adopt, should a standard gate array of such logic cells be required.

3 THE CHOICE OF THE BASIC ULG.2

The choice of the particular ULG candidate for standardisation will be considered, bearing in mind:

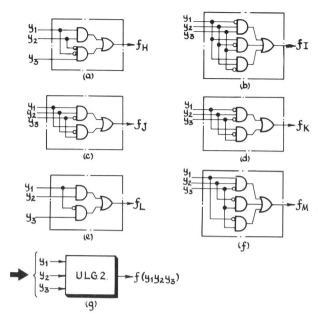

FIG. 3   The six basic ULG.2 candidates   (a) $f_H = \{y_1 y_2 + \bar{y}_1 \bar{y}_2 y_3\}$
(b) $f_I = \{\bar{y}_1 y_2 y_3 + y_1 \bar{y}_2 y_3 + y_1 y_2 \bar{y}_3\}$ (c) $f_J = \{y_1 y_2 \bar{y}_3 + \bar{y}_1 \bar{y}_2 y_3\}$
(d) $f_K = \{\bar{y}_1 y_2 \bar{y}_3 + y_1 \bar{y}_2 y_3\}$ (e) $f_L = \{y_1 y_2 + \bar{y}_1 y_3\}$
(f) $f_M = \{y_1 y_2 \bar{y}_3 + \bar{y}_2 y_3 + \bar{y}_1 y_3\}$ (g) general ULG.2 schematic.

(a) the number of ULGs which are required to realise a range of duties in comparison with conventional NAND gates,
(b) the simplicity of system design using the chosen ULG candidate, particularly for the educational field where computer-aided design (C.A.D.) software may be limited or indeed absent,

and

(c) the circuit complexity and silicon area per ULG circuit in comparison with, say, conventional NAND gates.

We will confine our discussions here to consideration of the ULG.2 candidates f_H, f_L and f_M, as previous work has shown that these represent the best initial selection.

A simple comparison of these three NP-complete candidates will show that f_H is the least attractive. If we compare its specification with that of f_L, namely

$$f_H = y_1 y_2 + \bar{y}_1 \bar{y}_2 y_3 \text{ and } f_L = y_1 y_2 + \bar{y}_1 y_3$$

it is seen that there is an additional variable in the internal product factor $\bar{y}_1 \bar{y}_2 y_3$ of f_H, which is clearly not essential from the universality of the circuit, and which introduces no significant advantage in the input dedication patterns.

Further, as will be noted later, there is no easy synthesis algorithm for functions of $n > 2$ using f_H as there is for functions f_L and f_M, and therefore we may conclude that the optimum choice for a ULG.2 cell lies between the two candidates f_L and f_M.

The input connections required to realise all $n \leqslant 2$ functions are given in Table 1 for the latter two cases. Note that several alternative input connection patterns may be possible for certain functions. If a particular ULG configuration is fabricated with any input negated from that shown in Fig. 3, then the appropriate input vector in Table 1 requires negation, whilst if the circuit output is negated from the algebraic expression above, then the appropriate pairs of input connection patterns, e.g. A and $\bar{A}$ entries, or AB and $\bar{A} + \bar{B}$ entries, etc., require interchanging. Once a specific ULG configuration has been adopted for standardisation, then, clearly, the input connection patterns required to realise a range of duties are part of the once-and-for-all documentation procedure for the chosen cell.

The number of ULG circuits required for a range of anticipated applications is a particularly vital parameter. The smaller the number necessary, the fewer the on-chip, inter-cell connections and hence the more efficient the final design and system realisation. It may be thought that, given a wide range of applications, the average number of ULG.2 circuits would be identical, irrespective

TABLE 1 *Input connection patterns to realise all $n \leqslant 2$ combinatorial functions, using the NP-complete ULG.2 candidates f_L and f_M. Note that alternative input patterns are possible, including don't-care input connections in certain cases.*

$n \leqslant 2$ Function $f(x)$	Possible input connection pattern					
	f_L			f_M		
	y_1	y_2	y_3	y_1	y_2	y_3
0	0	0	0	0	0	0
1	1	1	0	0	0	1
A	A	1	0	A	1	0
$\bar{A}$	A	0	1	A	1	1
B	B	1	0	B	1	0
$\bar{B}$	B	0	1	B	1	1
AB	A	B	0	A	B	0
A$\bar{B}$	B	0	A	A	B	A
$\bar{A}$B	A	0	B	B	A	B
$\bar{A}\bar{B}$	A	0	$\bar{B}$	$\bar{A}$	$\bar{B}$	0
A + B	A	A	B	$\bar{A}$	$\bar{B}$	1
A + $\bar{B}$	B	A	1	$\bar{A}$	B	1
$\bar{A}$ + B	A	B	1	A	$\bar{B}$	1
$\bar{A}$ + $\bar{B}$	A	$\bar{B}$	1	A	B	1
A $\oplus$ B	A	$\bar{B}$	B	1	A	B
A $\bar{\oplus}$ B	A	B	$\bar{B}$	1	A	$\bar{B}$

of what variant of ULG.2 circuit was chosen. This does not, however, occur.

By way of illustration, let us consider the number of ULG.2 circuits required to realise all possible functions of ≤ 3 variables, that is 256 combinatorial functions, using the two ULG.2 candidates f_L and f_M. When input and output negations on f_L and f_M are introduced, it is found that there are four functional variants of f_L, and six functional variants of f_M. Several circuit variants of each are functionally identical, e.g. circuit $y_1 y_2 + \bar{y}_1 y_3$ is functionally identical to circuit $\bar{y}_1 y_2 + y_1 y_3$, as this is merely an interchange of terminal designations. Thus, we have a total of 10 ULG.2 variants in the statistics shown in Table 2, each being a 3-input, 1-output circuit, which has the merit of being directly comparable with previously-published statistics involving the number of 3-input NAND gates and 3-input NOR gates required for this same range of duties[12].

It may be observed from Table 2 that a three-input ULG.2 is approximately twice as powerful as a 3-input NAND or NOR gate in general applications. Thus the number of inter-cell connections in a master-gate array will be approximately half if 3-input ULG.2 gates are chosen in comparison with 3-input NAND/NOR gates. However there is no overriding advantage of f_L over f_M or vice versa from Table 2.

This leaves us with three further factors to consider, which may individually or collectively indicate a preferred choice. These are:
(i) the relative ease of design of functions of $>n$ input variables with the ULG.2 cells f_L and f_M,

TABLE 2 *Statistics relating to 3-input circuits in the realisation of all 256 functions of $n \leq 3$. NAND, NOR results taken from Hellerman. Note that our original ULG.2 expression for f_M has been refactorised as $\{y_1 y_2 \oplus y_3\}$.*

Type of circuit, 3-input NP-complete ULG.2 or Boolean gate	Total no. necessary to realise all 256 functions	Average no. per function	Average no. of circuits in cascade per function
$f_L, \{y_1 y_2 + \bar{y}_1 y_3\}$	611	2.39	1.93
$f_L, \{y_1 \bar{y}_2 + \bar{y}_1 y_3\}$	609	2.38	1.87
$f_L, \{y_1 y_2 + \bar{y}_1 \bar{y}_3\}$	609	2.38	1.87
$f_L, \{y_1 \bar{y}_2 + \bar{y}_1 \bar{y}_3\}$	577	2.25	1.93
$f_M, \{y_1 y_2 \oplus y_3\}$	661	2.58	1.93
$f_M, \{y_1 y_2 \oplus \bar{y}_3\}$	625	2.44	1.88
$f_M, \{\bar{y}_1 \bar{y}_2 \oplus y_3\}$	661	2.58	1.93
$f_M, \{\bar{y}_1 \bar{y}_2 \oplus \bar{y}_3\}$	625	2.44	1.88
$f_M, \{\bar{y}_1 y_2 \oplus y_3\}$	540	2.11	1.83
$f_M, \{\bar{y}_1 y_2 \oplus \bar{y}_3\}$	525	2.05	1.83
3-input NAND	1118	4.37	3.01
3-input NOR	1118	4.37	3.01

(ii) the silicon area required to fabricate f_L and f_M cells in any given technology, and
(iii) the ability of the two candidates to be used in storage and sequential applications.

It may be shown that considerations (i) and (ii) above do not greatly favour f_L over f_M or vice versa;* in particular, synthesis algorithms are available, both with the multiplexer-type structure of f_L and with the Reed-Muller structure of f_M[8], unlike all the other alternative candidates shown in Fig. 3 which do not lend themselves to any known synthesis procedure. However when the third and final factor above is considered, then candidate f_L shows particular advantages, as will be briefly noted.

For a ULG cell to be useful in sequential applications, the ready realisation of basic storage elements, such as RS latch and clocked D-type circuits, is desirable, ideally using only one ULG cell. Clearly, since a ULG circuit is capable of realising a basic NAND (or NOR) logic gate, it is always possible to fabricate storage requirements using each ULG circuit as a NAND (or NOR) element, but this is wasteful of the full logical power of the ULG cell. However, the ULG.2 circuit f_L is, by itself, capable of realising a number of storage requirements, but there is *no variant* of f_M which has this capability; additional logic is always required with f_M to complete the necessary positive feedback loop. The various variants on f_L, however, have the various capabilities shown in Fig. 4.

4 THE ULG.2 CELL FOR EDUCATIONAL PURPOSES

Because clocked D-type elements are supremely significant in many digital system designs, enabling data to be clocked into or out from a combinatorial sub-system, or to provide shift-register action, it is considered that this should be a dominant consideration in the choice of ULG.2. This leads us to candidate f_L rather than f_M, even though f_M is marginally more powerful in combinatorial applications (see Table 2).

Within the variants of f_L, we initially discount the fourth, see Fig. 4, for the reason that this one cannot be used by itself for delaying action. This leaves the first three variants, whose storage attributes per circuit can be summarised as shown in Table 3. Unfortunately, none of these are ideal, i.e. RS with output Q, and also D-type with output $Q_{n+1} = D$.

However, for educational purposes, where the 'breadboard' chip may be used by a variety of persons with a spread of design abilities, it is considered essential to have textbook RS and D-type action readily available, rather than any complementary action; this may not be necessary for an industrial version, where familiarity will be available. Thus if we here set our target to give

(i) textbook RS latch, and
(ii) clocked type D, clocked on clock high,

---

*Note that, for conformity, all the candidates in Fig. 3 are shown in AND/OR form. Candidate f_M may also be expressed as $f_M = y_1 y_2 \oplus y_3$, and as such has a silicon area comparable to f_L.

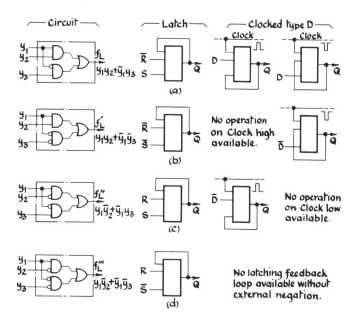

FIG. 4   *The storage possibilities which are directly available from the four functional variants of ULG circuit f_L, see Table 2.*

TABLE 3

	Latch	Clocked D-type
f_L:	makes $\overline{RS}$ with output Q	makes D-type with output Q, (i.e. $Q_{n+1} = D$), clocked on clock high or clock low
f'_L:	makes $\overline{RS}$ with output Q	makes $\bar{D}$-type with output Q, (i.e. $Q_{n+1} = \bar{D}$), clocked on clock low
f''_L:	makes RS with output Q	makes $\bar{D}$-type with output Q, (i.e. $Q_{n+1} = \bar{D}$), clocked on clock high

from one ULG.2 cell, then we must modify our previously chosen candidate f_L as follows. Note that the f_M variants remain inconvenient for general-purpose use, since, for combinatorial purposes, they involve Exclusive relationships with which many people are unfamiliar.

Thus, we require some addition to the f_L variants in order to provide our desired RS and D-type capabilities, this addition being either a further input, or an additional complementary output. All four variants of f_L will now be found to give these objectives with such an additional input or output connection.

Fig. 5 illustrates these possibilities. Note that because the first and fourth variant, and the second and third variant of f_L are complements of each other, see Fig. 4, then if two outputs Q and $\bar{Q}$ are provided per ULG cell there is no functional difference whatsoever between variants f_L and f'''_L and between f'_L

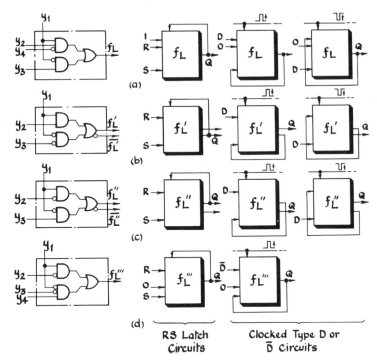

FIG. 5   Additions to the four f_L variants of Fig. 4 in order to give true RS and D or $\bar{D}$-type action using one ULG cell.
(a) input addition to f_L (b) output addition to f'_L
(c) output addition to f''_L (d) input addition to f'''_L
  (Note that $Q_{n+1} = \bar{D}$ with circuit (d); $Q_{n+1} = D$ not available).

and f''_L. Hence the choice narrows down to the 4-input/1-output candidate of Fig. 5(a), or the 3-input/2-output candidate of Fig. 5(b).

Both of these ultimate candidates have considerable merit. To a large extent the final question is:

Is a 4-input/1-output ULG cell more useful than a 3-input/2-output ULG cell, where true and complemented outputs are always available from the latter?

For educational purposes it is recommended that the *first candidate* be adopted, particularly because its action is simple and straightforward to appreciate, without having to be concerned with which one of two outputs to use for different duties. It also has the merit of being able to use its 3-input AND gate to realise directly a number of combinatorial functions of $n > 2$ which are beyond the capability of the 3-input/2-output ULG.2 variants. Our final choice for general educational purposes therefore is as illustrated in Fig. 6(a).

An input programming schedule for this ULG cell to realise any non-trivial combinatorial function of ≤ 2 input variables A, B is shown in Table 4.

Alternative input connection programming patterns are possible for many of

TABLE 4

Function	Inputs			
	y_1	y_2	y_3	y_4
$\bar{A}$	1	1	dc	A
$\bar{B}$	1	1	dc	B
AB	A	B	0	0
$\bar{A}B$	1	B	0	A
$A\bar{B}$	1	A	0	B
$\bar{A}\bar{B}$	A	0	$\bar{B}$	dc
$A+B$	A	1	B	0
$\bar{A}+B$	A	B	1	0
$A+\bar{B}$	B	A	1	0
$\bar{A}+\bar{B}$	A	1	1	B
$A \oplus B$	A	1	B	B
$A \oplus B$	A	B	$\bar{B}$	0

these functions. Additional combinatorial functions of >2 input variables which are also directly available from this ULG.2 cell include:

$AB\bar{C}, A\bar{B}C, \bar{A}BC$

$AB\bar{C} + \bar{A}$ and permutations

$A\bar{B} + \bar{A}C$ and permutations

$AB + AC$ and permutations

$AB\bar{C} + A\bar{B}$ and permutations

A library of the required input connections to realise a complete range of functions may be readily compiled and held on file.

For storage requirements, the latch and type D configurations illustrated in Fig. 6(b) to (g) are immediately available. More complex type JK assemblies etc. may readily be compiled as a once-and-for-all design exercise.

The use of this cell in combined combinatorial and sequential applications readily follows. Fig. 7 illustrates possible structures which may be used for micro-instruction and similar clocked-data-in/clocked-data-out system applications.

5 MASTER-CHIP L.S.I. FABRICATION TECHNOLOGY

It is particularly essential for educational and many R. and D. purposes to be able to command a rapid production schedule of circuits designed by students or research staff. The gate array approach will aid this objective. However, the technology employed in the circuit fabrication should be well established, and not the latest, unproven, frontier of the art. Both of these objectives will be met if a standard CMOS production technology is chosen; in particular IsoCMOS™ is rapidly becoming the standardised technology for several new I.C. production lines in the U.K.

Thus, the fabrication of this ULG gate array should be based upon CMOS technology, representing as it does the technology which will dominate all but

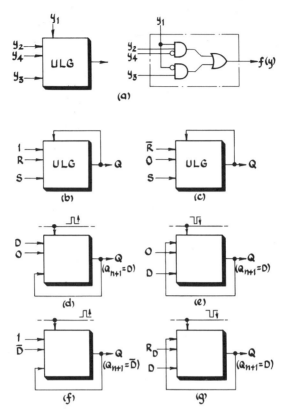

*FIG. 6 The chosen ULG.2 and its storage connections
(a) the 4-input/1-output circuit, $f(y) = \{y_1 y_2 \bar{y}_4 + \bar{y}_1 y_3\}$ (b) normal RS latch
(c) $\bar{R}S$ latch (d) clocked type D, transfers D to output on clock high (e) clocked
type D, transfers D to output on clock low (f) clocked type $\bar{D}$, transfers $\bar{D}$ to
output on clock high (g) clocked type D with direct reset, transfers D to output
on clock low.*

the high-speed computer market in the 1980s. Preliminary silicon layouts for the ULG.2 cell shown in Fig. 6 indicates that with 5 micron geometry and generous design rules, a gate array as shown in Fig. 8 would have the following principal characteristics:
(i)   ULG.2 gate size 80 × 100 microns
(ii) max. clock speed 5 MHz
(iii) chip size for 15 × 15 gate array (225 ULG cells) with 40 I/O buffer circuits 3.3 × 3.3 mm
(iv) ditto for 20 × 20 gate array (400 ULG cells) 3.8 × 3.8 mm
These parameters are well within the established state of the art.

A CMOS breadboard gate array

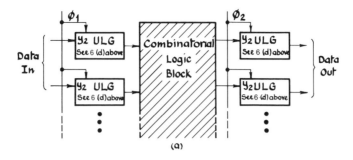

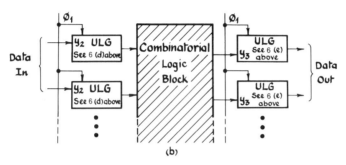

FIG. 7   Use of ULG.2 cells in synchronous logic networks (a) two-phase clocking; data clocked in on ϕ_1 high, data clocked out on ϕ_2 high (b) single phase clocking; data clocked in on ϕ_1 high, data clocked out on ϕ_1 low.

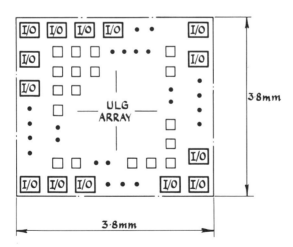

FIG. 8   The uncommitted master chip gate array, containing a matrix of fully-functional ULG cells (Note that the dimensions shown are provisional for a $20 \times 20 = 400$ cell array).

6 CONCLUSIONS

We have here discussed the formulation of a master-chip L.S.I. circuit, which has particular attributes as a vehicle for the practical implementation of student designs, performed during L.S.I. design courses and similar educational activities. The availability of fully-functional cells on the master-chip eliminates the necessity for a student or research person to be concerned with design details down to individual component level, and leaves him free to concentrate upon the functional interconnection part of his activities, whether this be a system design exercise, or a C.A.D. software design exercise, or both.

The fundamental universality of the ULG cell is found to be particularly attractive for rapid design purposes. Any fears of unfamiliarity in comparison with, say, NAND or NOR gates is rapidly dispelled. The essential feature is that any given logic circuit may be immediately partitioned into 'boxes' corresponding to the capabilities of the universal cell, without the complication of having to generate all-NAND or all-NOR equivalent circuit realisations. Similarly, any Boolean logic expression can be factorised and directly implemented, again without any conversion to all-NAND or other restricted form. This procedure closely mirrors what many designers now do when using discrete SSI packages for a circuit implementation, the ULG merely requiring initial documentation to give the designer the 'pin connections' necessary to realise its various logic capabilities.

ULG gate-arrays may therefore be considered as ideal L.S.I. breadboard circuits, requiring only the interconnect masks to convert them into custom-designed functional entities. In spite of current restraints in research budgets for new ventures, it is hoped that fabrication plans for a ULG master-chip may be forthcoming in the near future, so that uncommitted chips and design commitment data may be made available to U.K. educational establishments.

7 ACKNOWLEDGEMENTS

Research work at the University of Bath on universal logic gates is funded by the Wolfson Foundation, Wolfson Grant ref. Z/80/3. This work includes the comprehensive analysis of all ULG circuits and gate arrays, but does not cover fabrication for educational purposes. Funding for the latter area is being sought.

8 REFERENCES

[1] I.C. Cost Consultants Inc., *LSI gate arrays, Market Study Report* (Nov. 1979).
[2] Twaddell, W. 'Uncommitted I.C. logic', *EDN*, pp. 89–98 (April 5, 1980).
[3] Hartmann, R. F. 'Design and market potential for gate arrays', *Lambda*, **1**, No. 3, pp. 55–59 (1980).
[4] Special Report, 'Gate arrays', *Electronics*, pp. 145–158 (Sept. 1980).
[5] Rehman, M. A., 'Custom-integrated circuits', *Electronic Engineering*, **52**, pp. 55–68 (April 1980).
[6] Yau, S. S. and Orsic, M., 'Universal logic modules', *Proc. 3rd Princeton Conf. Information Sci. Sys.*, pp. 499–502 (March 1969).
[7] Preparata, F. P., 'On the design of universal Boolean functions', *Trans. IEEE*, **C.20**, pp. 418–423 (1971).

[8] Hurst, S. L., *Logical Processing of Digital Signals*, Crane-Russak, NY, and Edward Arnold, London (1978).
[9] Edwards, C. R., 'A special class of universal logic gate and its evaluation under the Walsh transform', *Int. J. Electronics*, **44**, pp. 49–59 (1978).
[10] Hurst, S. L., 'Custom LSI design: the universal-logic-module approach', *Proc. IEEE Int. Conf. on Circuits and Computers*, pp. 1116–1119 (1980).
[11] Chen, X. and Hurst, S. L., 'A consideration of the minimum number of input terminals on universal-logic-gates, and their realisation', *Int. J. Electronics*, **50**, pp. 1–13 (1981).
[12] Hellerman, L., 'A catalogue of 3-variable OR-invert logic circuits', *Trans. IEEE*, **EC12**, pp. 198–223 (1961).

4

A PRACTICAL APPROACH TO DIGITAL INTEGRATED CIRCUIT DESIGN USING UNCOMMITTED LOGIC ARRAYS

P. J. HICKS
Department of Electrical Engineering and Electronics, University of Manchester Institute of Science and Technology, England

INTRODUCTION
With microelectronics playing an increasingly important role in many fields, and particularly in the areas of computer engineering and information technology, there is a growing demand for engineers equipped with the skills needed to design integrated circuits and systems. The provision of practical training to support taught courses on integrated circuit (IC) design, while highly desirable, can pose problems for higher educational establishments such as universities and polytechnics in view of the expensive facilities required to produce microelectronic devices.

Several accounts have been written[1,2] describing how suitable equipment for small-scale IC manufacture can be assembled on a limited budget. Although these laboratories have proved to be very successful in introducing students to the basic wafer processing steps associated with silicon IC fabrication such as photolithography, oxidation, diffusion etc., they are unable to serve as production lines for large scale (LSI) designs containing, maybe, several thousand transistors. Because of the very high costs of setting up and running an LSI microfabrication facility, this role must be confined to a small number of higher education establishments or, alternatively, left to industry.

The proposition has already been put forward that the teaching of IC design is considerably enhanced by students being able to design their own integrated circuits and subsequently have the chips fabricated and returned to them for characterisation and testing. It is not difficult to justify this notion by making a simple analogy between IC design and computer programming. Many people would probably agree that teaching computer programming is a fruitless exercise unless students are given the opportunity to write their own programs. Having developed a new piece of software, it is imperative that it should be run on a computer, simply in order to reveal any errors that it may contain and to see if it functions as originally intended. Only in this way, with suitable guidance and supervision, can mistakes be corrected and sound programming skills be developed.

Running an IC design through a wafer fabrication line can be likened to running a program on a computer. Just as there are sets of design rules that must be adhered to in IC design, so the syntax of a programming language imposes a similar set of constraints on the programmer. More recently, the pioneering work of Mead and Conway[3] in the teaching of VLSI (Very Large Scale Integration) systems, suggests that structured design methods are equally as important to the IC designer as they have already been shown to be in computer programming.

Although access to silicon IC processing for teaching purposes has certainly become easier and more widely available during the past few years[4, 5], the cost per student of providing a full custom IC fabrication service is still quite high. If IC design is to be widely taught to large numbers of students it is clearly worth investigating how this might be achieved more economically.

As a first step towards meeting the objectives outlined above, semi-custom ICs provide a simple and cost-effective introduction to the practical teaching of IC design. In contrast to full-custom design, where every detail of the circuit layout has to be specified on anything from half a dozen to a dozen different levels in order to define uniquely the characteristics, positioning and interconnection of the various components on the chip, one of the principal advantages of semi-custom ICs is that the designer is isolated from much of the complexity of low-level chip design. This is done by constraining him to use pre-defined components or groups of components laid out in a regular array, thus removing the necessity to be concerned with the characteristics of individual devices. There are several ways in which this can be achieved and two of them are summarised below:

(i) *Gate Arrays or Uncommitted Logic Arrays (ULAs)*

The concept on which all gate arrays are based is of a chip containing a regular matrix of logic gates or components, which is pre-processed up to the final stages of metal layer patterning. Approximately 90% of the wafer processing steps have therefore already been completed before the chips reach the stage where they will be committed to a particular function. This normally involves the patterning of a thin film of aluminium covering the surface of the chip to form the metal tracks which connect the components together. The pattern of tracks uniquely defines the circuit function and therefore 'customises' the chip to suit the desired application. In place of the many (typically 6 to 12) photolithographic masks that have to be uniquely designed in order to produce a full-custom chip, only one (or possibly three, if two layers of metal interconnection are used) has to be produced to 'wire up' the prefabricated array of components on a gate array. The amount of design effort required is thus considerably reduced, first of all because design can be carried out using fully characterised components at the logic level rather than at the device level, and secondly because much less time is taken up in the layout of the masks.

(ii) *Standard cell or library custom ICs.*

This technique exploits the hierarchy inherent in all logic systems by partitioning the logic into functional building blocks or 'standard cells'. The cells themselves are custom-designed and held in a cell library. For a given design, appropriate cells are selected from the library and placed in rows across the chip. Sufficient space is left between the rows to allow for interconnections between the cells. It is evident that the standard cell technique will typically require as many photolithographic masks as a full-custom design. However, because all the cells are pre-defined and, furthermore, are constrained to occupy positions in the rows, the process of placing and interconnecting the cells can usually be highly automated.

Reviews of the complete range of semi-custom ICs available can be found elsewhere[6,7]. Semi-custom analogue circuits can also be constructed from arrays of uncommitted components using a single metal layer masking step. The use of an array of this type for teaching purposes has previously been described by Poole[8].

Both of the digital semi-custom design techniques described above have one thing in common: because they substantially lessen the amount of time and effort needed to integrate a given logic system onto silicon when compared with a full-custom design, it follows that the cost of a semi-custom integration is correspondingly lower also. Whereas the development costs for a new LSI custom IC are so high that it is uneconomic to produce unless it can be manufactured and sold in large quantities ($>100,000$ units), it may be possible to design and produce an equivalent semi-custom device in a fraction of the time and at a cost which makes it economic in quantities of the order of a few thousand. Of the two techniques, the first is undoubtedly the cheapest when it comes to producing small quantities, since there is only a single customising mask step to be performed. In line with the objectives defined earlier, therefore, the Uncommitted Logic Array (ULA) has been chosen as a means of providing low-cost practical training in semi-custom IC design.

The remainder of this paper describes how the teaching of IC design based on the ULA has been incorporated into a Masters degree course at UMIST. The next section gives a brief description of the course and is followed by more detailed descriptions of the ULA itself and the facilities needed in order to have students' designs fabricated.

COURSE STRUCTURE

The UMIST M.Sc. course on Integrated Circuit System Design was established with the support of the U.K. Science Research Council (now the Science and Engineering Research Council — SERC) in 1980. Its object is to provide students with the knowledge and practical skills needed for the efficient design, manufacture and testing of components and systems based on microelectronics technology. The course includes a wide range of material spanning several different disciplines and consequently draws upon the resources of three academic departments, namely, Electrical Engineering and Electronics, Computation and Management Sciences.

The course begins in October each year and is of one year duration. It follows a fairly conventional pattern, consisting of lectures and laboratory work during the first two terms from October through to April followed by a research project carried out between May and the following October.

The subject areas taught on the course cover the entire hierarchy of activities associated with IC design. Beginning at the lowest level, lectures are given on semiconductor device models, silicon wafer processing, IC fabrication techniques and IC circuit theory. The design of integrated circuits and systems, logic design and the structure of microelectronic systems are all dealt with at the intermediate level, while at the highest level the treatment expands to include software engineering, programming languages and design automation techniques.

A great deal of emphasis in the course is placed on giving students practical experience of integrated circuit design. The design activities are based on the semi-custom approach referred to earlier, on the one hand, and full custom design using NMOS technology on the other. ULA design is practised in the first term of the course and full-custom NMOS design in the second. This arrangement allows sufficient theory of full-custom design techniques to be taught in Term 1 to provide a sound basis for the practical work in Term 2. In contrast, a basic knowledge of transistor circuit theory and logic design seem to form an adequate background for the ULA design work. Apart from these activities, time is also spent in the silicon wafer processing laboratory where experience is gained of the various stages of IC manufacture — photolithography, oxidation, diffusion, assembly and packaging etc. Other laboratories provide practice in logic design, the application and programming of microprocessors and the structured design of programs using a high-level language.

The project phase of the course culminates with the submission of a dissertation describing the work carried out by the student. Projects may be undertaken in any of the diverse subject areas covered by the course and are supervised by a member of staff.

Although the sections that follow deal specifically with the practical teaching of IC design based on the ULA, many of the facilities to be described are also used for the full-custom NMOS design work. A full account of the latter will be presented at a later date.

THE UNCOMMITTED LOGIC ARRAY

As previously explained in the Introduction, an Uncommitted Logic Array consists simply of a large number of unconnected components (mainly transistors and resistors) that are prefabricated in a regular matrix arrangement on a silicon substrate. By interconnecting the components in a particular way, any arbitrary logic function can be integrated onto a single chip, within the limit imposed by the total number of logic gates available. The pattern of interconnections is etched into a thin film of metal (usually aluminium) which constitutes the uppermost layer of the chip. The process of committing the device to implement a given function, therefore, involves a single lithographic step in which the pattern of tracks for interconnecting the components is transferred from a photographic plate called a mask to the metal layer.

The range of ULAs manufactured by Ferranti Electronics Ltd. extends from the small (150 gates) to the very large (2000 gates). They can be optimised either for low-power dissipation, claimed to be comparable to that of complementary MOS (CMOS) technologies, or for high-speed performance, matching Schottky TTL figures. In addition, some of the chips offer the capability of integrating both digital and analogue functions onto the same chip, thus further increasing their potential for finding 'single-chip' solutions to certain design problems.

All of the Ferranti ULAs are fabricated using a bipolar technology known as Collector Diffusion Isolation or CDI[9]. Compared to conventional junction-isolated bipolar transistors, devices fabricated in CDI technology consume far less surface area on the chip because the collector diffusion also serves to isolate the device from its surroundings. With minimum feature sizes of the order of 4 to 5 µm this technology is easily capable of integrating some 2000 gates onto a single chip and therefore qualifies as a bipolar LSI technology.

The particular ULA which has been used at UMIST to support the teaching on the M.Sc. course comprises 440 cells, each of which can be connected to yield a pair of two-input NOR gates. A diagram of the ULA cell in plan view is shown in Fig. 1(a) and, beneath it, in Fig. 1(b) is the circuit schematic for one of the pair of NOR gates it can produce. The array cells are fully characterised in terms of their electrical parameters and switching performance in a manual supplied by the manufacturer[10]. A selection of the major array characteristics are listed in Table 1.

The manual also gives a description of the ULA development cycle and detailed instructions for the design engineer responsible for the layout of the interconnections.

Around the borders of the array are located the bonding pads that will ultimately be used to connect the chip to its package pins. Associated with each bonding pad is a peripheral cell containing a further group of uncommitted components which, in this case, are used to interface the logic on the array cells with the outside world. All these details can be observed in the photograph shown in Fig. 2 of a ULA taken before it has been committed. Inputs and outputs may be chosen by the designer from a library of standard peripheral cells to suit the requirements of TTL and CMOS logic as well as providing more sophisticated functions such as Schmitt triggers, oscillators and even analogue comparators.

One particularly attractive feature of the Ferranti ULA is the manner in which power is distributed to the cells in the array. Instead of having to give up valuable metal interconnection area to carry the $+0.9$ V power rail and ground around the chip, this task is performed in the bulk of the silicon itself. The positive supply is carried in a 'sea' of $n+$ diffused silicon that surrounds all the array cells while the ground return path is provided by the p-type substrate. Contact holes through to each of the power supply 'rails' are located within each cell of the array, as shown in Fig. 1(a).

Digital integrated circuit design using uncommitted logic arrays

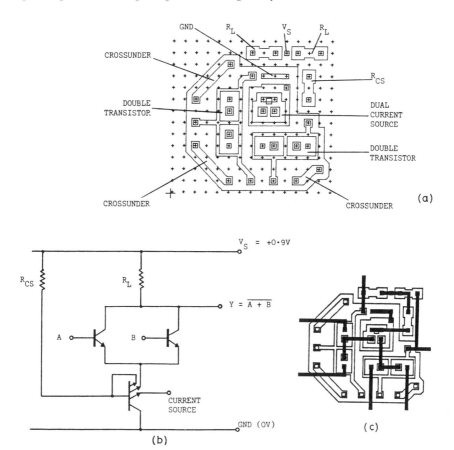

FIG. 1 (a) The layout of the components within one cell of a 440 cell ULA, (b) The circuit schematic is drawn for one of the pair of Current Mode Logic (CML) gates that can be produced from the cell components, (c) A typical configuration of interconnecting tracks needed to produce two of the gates in (b).

TABLE 1 Major electrical and switching characteristics of the Ferranti 440 cell ULAs.

	440 Cell ULAs	
	ULA 5C000	ULA 5N000
Gate delay	9 ns	25 ns
Gate power	0.25 mW	70 μW
Clock rate	20 MHz	6 MHz
V_{cc}	5.0 V	5.0 V

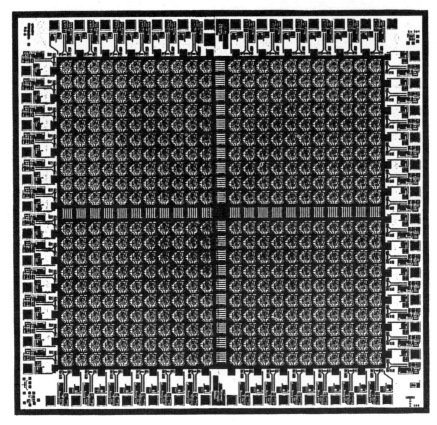

FIG. 2 The Ferranti 440 cell ULA before the metallisation layer has been customised. The bonding pads and their associated peripheral cells can be clearly seen surrounding the four quadrants which contain the array cells.

COMPUTER-AIDED DESIGN TOOLS

One of the most important factors that has led to the increased popularity of semi-custom ICs is the wider availability of computer-aided design (CAD) tools to ease the burden on the designer[11]. Indeed, some manufacturers now claim to be able to offer almost fully automated design systems for certain types of semi-custom IC. The input to such a system would, typically, be at the logic diagram level and the output produced would be the layout for the mask or masks needed to fabricate the device. A set of test pattern sequences that will enable the completed devices to be functionally tested would probably also be produced.

As yet a full Design Automation (DA) system capable of taking a design all the way from logic diagram to layout is not available for ULAs, although progress is being made in that direction[12]. There are several reasons for this.

One is that automated layout of interconnections is considerably easier for a computer to do if it has two levels available on which to run the tracks. Secondly, it is significant that most semi-custom ICs which claim to be amenable to full DA only achieve this goal at the expense of a somewhat lower gate packing density compared to arrays like the ULA that rely on manual layout techniques.

The CAD tools that are available at UMIST are different and less comprehensively geared to the needs of ULA design than those used by Ferranti themselves. By virtue of the support given to the M.Sc. course by the SERC, UMIST has access to the GAELIC suite of programs for IC design. GAELIC, which stands for Graphic Aided Engineering Layout of Integrated Circuits, was originally developed by the University of Edinburgh and is now marketed by Compeda Ltd. It provides numerous facilities to assist the designer of integrated circuits, including:

(i) Entry of graphical data to a computer — three methods are available:
    (a) GAELIC language — a graphics description language which is input to the computer as a text file.
    (b) Input from a co-ordinate digitiser.
    (c) Direct input via a graphics display terminal.
(ii) Graphical editing.
(iii) Plotting.
(iv) Design rule checking (DRC).
(v) Generation of pattern generator drive tapes (used for making mask plates).
(vi) Interfaces to other graphics systems, e.g. CALMA, APPLICON Computervision etc.
(vii) Logic simulation.
(viii) Automatic IC layout.

Some of the programs listed above are intended primarily for full-custom rather than semi-custom IC design. For example, the design rule checker and automatic IC layout software are irrelevant in the present context. The programs used at UMIST during the preparation of the ULA designs are (i) (b), (ii), (iii), (vi) and (vii). These and the other GAELIC programs are all resident on a PRIME 750 computer situated at the SERC's Central Microfabrication Facilities. The terminals in the IC design laboratory at UMIST are linked to the PRIME via a 2400 baud serial line. A schematic diagram illustrating the various terminals and their connections into the SERC computing network is shown in Fig. 3. The visual display units (VDUs) are principally used for gaining access to the logic simulator while the graphics display terminals (Tektronix type 4014 storage tube graphics displays) are used in conjunction with the digitiser for data entry, or on their own for graphical editing. Checkplots up to A3 paper size can be obtained in the laboratory on four-colour graphical X-Y plotters (Hewlett-Packard type 7221B). Although the GAELIC programs are run on the PRIME computer at the Rutherford Appleton Laboratory, files can be readily transferred across the network to a

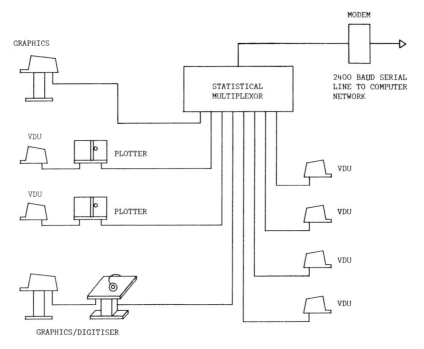

FIG. 3 A schematic diagram showing how the various terminals in the IC Design Laboratory are connected to the PRIME 750 computer at the SERC's Rutherford Appleton Laboratory via the computing network.

similar machine on the UMIST campus. In this way use can be made of a larger (42-inch) drum plotter and files can also be written locally onto magnetic tape. This latter facility provides the means for transporting the ULA design data from UMIST to the APPLICON-based CAD system used by Ferranti at their Microelectronics Centre in Hollinwood, Lancashire. Once this stage has been reached, Ferranti's engineers add the layout details for the peripheral cells and complete a series of final checks before going on to manufacture the mask for the metal interconnection layer.

THE ULA PROJECTS

The ULA design projects are carried out during the first term of the M.Sc. course, in the ten weeks from the beginning of October to the middle of December. As an introduction to designing logic on the ULA, the students begin with a series of exercises which progress from simple gates through to the design of a multi-stage counter circuit. Following standard practice, the layout of the interconnecting tracks is carried out manually by drawing them in as lines with coloured pencils on a sheet of mylar-based draughting film. This is superimposed on a master layout sheet supplied by Ferranti which shows the details of all the cells in the array at 240 × full size. Also provided on the master layout sheet, is a grid corresponding to 0.5 mil (~12.5 µm) intervals on the finished chip. The grid positions in a single cell can be seen in Fig. 1(a), where it will also be noticed that all the contact holes to the cell components are centred at grid points. One of the basic design rules for laying out the tracks on the ULA is that they must always start and finish on a contact hole and, in between, they must follow the grid lines.

The preliminary exercises mentioned above familiarise the students with the basic principles of ULA design and rapidly build up their confidence to the point where they are able to undertake the design of a logic subsystem of the order of 100 to 200 gates. Since the 440 cell ULA is conveniently arranged as four quadrants with 110 cells in each, every student on the course is allocated one of these quadrants for his or her design. If two students work together on a larger design, they share half of a ULA between them, each taking responsibility for the layout of one quadrant. For teaching purposes, this scheme clearly has the advantage that the wafer fabrication costs are significantly reduced since a class of, say, 16 students can be accommodated on four different ULAs. There is one potential drawback to this scheme, which arises from the fact that each quadrant design can only use a maximum of 12 or 13 bonding pads, this being the number in the immediate vicinity of each quadrant. In practice, however, this limitation rarely constitutes a problem.

During the logic design phase, the students are urged to consider the problems of testing their circuits when the finished chips are returned. This is especially important, bearing in mind that the chips cannot be debugged by probing the internal circuitry with an oscilloscope or other test instrument. Apart from the microscopic size of the tracks, the load imposed on the gates by the relatively massive capacitance of the test probe would almost certainly cause the circuit to cease to function. Where necessary, therefore, extra bond pads must be utilised in order to permit access to strategically-located internal nodes so that signals may either be monitored or injected for testing purposes.

Design verification is carried out using the GAELIC logic simulator and gives students a high degree of confidence in their designs before starting on the layout phase. It also helps in identifying potential testing problems as described in the previous paragraph. Typical figures for gate propagation delays under various conditions of fan-in and fan-out are obtained from the ULA manual and input as timing parameters to the simulator.

Layout commences by partitioning the logic into modules and estimating the number of cells that will be needed to implement each module. The logic is then mapped onto the array by assigning appropriate groups of cells to each module and, finally, pencilling in the connections over the master layout sheet (Fig. 4).

When the layout is complete, it is carefully checked both by the students themselves and by members of staff. After any errors have been corrected, the track centre-line co-ordinates are interactively input to the PRIME computer using the digitiser table, graphics display and GAELIC digitiser program. A checkplot is next obtained from the 42-inch drum plotter on the same scale as the master layout sheet (i.e. 240 × actual size), and this is compared with the

FIG. 4 One of the students from the course can be seen here drawing in the pencil interconnections on a master layout sheet for one quadrant of a 440 cell ULA.

hand-drawn layout by superimposing one on top of the other. Any discrepancies at this stage, for example, due to errors during digitising, can be easily rectified with the GAELIC graphics editor. Only when the file containing the layout data is judged to be completely error-free is it ready to be translated from GAELIC format into APPLICON format. The APPLICON file is written onto magnetic tape for transportation to Ferranti's Microelectronics Centre where the final stages mentioned at the end of the previous section are performed. Before the metal layer mask plate is actually made, another checkplot is produced, this time by Ferranti. On this are shown all the tracks as full-width lines together with the bonding pads, peripheral cells and details such as alignment marks and processing test structures. These now appear exactly as they will be produced in the metal layer on the finished ULA. A thorough check is therefore made of all the layout information on this plot, since it represents the last opportunity for correcting any mistakes that may have escaped unnoticed in the earlier stages.

PACKAGING AND FINAL ASSEMBLY

After Ferranti have made the metal layer mask, it is used to process a number of the ULA slices. The complete wafers are then shipped to UMIST for final assembly into 40 pin ceramic dual-in-line packages. The sequence of steps required to accomplish this is briefly as follows. First of all the individual chips are probe-tested while the wafer is still intact in order to identify those that are functional from those that are not. The tests carried out at this stage are relatively crude since the object is merely to establish whether the chips are 'go' or 'no go'. A diamond-tipped scribing tool is then run down the gaps between the rows of chips on the wafer so that when gentle pressure is applied it fractures along the scribe lines to yield the individual chips (die). The good chips are bonded into packages with a silver-loaded epoxy resin following which the bonding pads are connected to the package pins with very fine gold wires using a thermosonic bonder.

All of the steps described above are carried out under clean room conditions in the laboratories of the Solid State Electronics Group in the Department of Electrical Engineering and Electronics. These facilities have been described in greater detail elsewhere in this issue[2].

TESTING

Apart from the very high level of motivation that it induces amongst the students, one of the most important reasons for getting the ULAs fabricated is that this is the only way their designs can be tested to see whether or not they actually work. Success at this stage has been seen to markedly increase the students' confidence in their ability as IC designers.

To facilitate the testing of the packaged ULAs, a comprehensive range of test equipment is available. The principal items are a 16-channel logic pattern generator (Hewlett-Packard type 8170A), an 8-channel word generator (Hewlett-Packard type 8016A) and a 16+8 channel logic state and timing analyser (Hewlett-Packard type 1615A). Functional testing of the ULAs can be carried out by generating input test sequences with the logic pattern generator while monitoring the circuit responses on the output pins with the logic analyser. By the time the chips are returned to the students, the test sequences that will be required should have already been determined, based on the earlier results obtained from the logic simulator.

At present the test pattern sequences have to be manually loaded into the pattern generator and, similarly, the results from the logic analyser have to be interpreted manually. It is planned that this process should be automated by connecting the three instruments to a small computer via the IEEE 488 General-Purpose Interface Bus (GPIB). A suitable minicomputer is available in the form of an LSI-11/23 processor (manufactured by the Digital Equipment Corporation) equipped with 128K bytes of main memory, a dual floppy disk storage system, graphics-enhanced VDU and GPIB interface. The operating system used at present is RT-11 and a PASCAL compiler is available, the code from which can be linked to the subroutines needed to drive the GPIB interface. Programs written in PASCAL will eventually enable an operator to specify the input test pattern sequences he requires, set up the instruments in the desired mode and subsequently analyse the test data obtained.

CONCLUSIONS

At the time of writing, the M.Sc. course has been in operation for two years. During the first year (1980–81) six students completed designs on two ULAs — these included a clock-calendar chip, a quad-multiplexed programmable baud rate generator for serial RS232 interfaces and the logic necessary for controlling requests for peripherals from a multi-microprocessor system.

The design phase, including logic design, simulation and layout, occupied the period from October to December. By mid-January the designs had all been checked, digitised and plotted and were ready for transfer to Ferranti. There was a delay of several weeks at this stage, owing to initial difficulties with the formatting of data on the magnetic tape. Once this problem had been solved, the transfer of data between the GAELIC and APPLICON systems became a routine process and in the current year (1981–82) was completed within a matter of days. The two metal layer masks were ready approximately a month later and the slices bearing the ULAs were delivered in April. The students were able to test their packaged devices during the summer term.

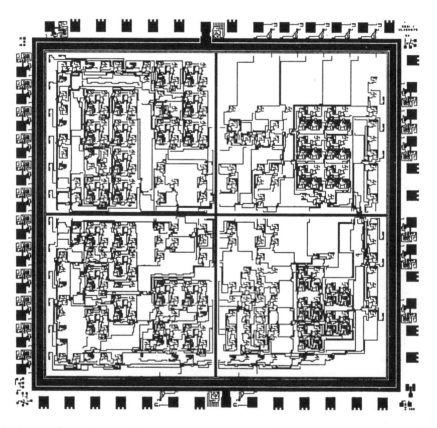

FIG. 5   The ULA metallisation pattern for the clock-calendar chip referred to in the text.

All of the designs produced by the students were found to work, although in some cases they did not function 100% according to specification due to minor design errors. The level of enthusiasm and commitment shown by the students in carrying out their ULA designs can only be described as astonishing and, in the author's opinion, fully justifies the time, effort and resources that have been invested in this project. In the present year (1981–82) ten students are registered on the course and, spurred on by the successes of the previous year, have already completed their designs on a total of three ULAs.

Even with the collaboration of a friendly ULA manufacturer like Ferranti, the cost of providing this service to the students is by no means small (although it is lower than the cost of full-custom design!). Apart from the cost of making the metal layer masks and processing the ULA slices, there are a number of 'hidden' costs such as the use of GAELIC, computer time and so on. In striving to reduce the cost per student even further, one possibility that is being investigated is to see whether the etching of the metal layer on the uncommitted slices could be carried out using the semiconductor processing clean room facilities at UMIST. Ferranti would continue to take the layout data and produce the metal layer mask, as at present, but then the mask and the uncommitted slices would be transferred to UMIST for the final processing stage. All the equipment required to do this, including a 3-inch mask aligner, is currently available[2].

An even more cost-effective method would involve the masks being fabricated using the Electron Beam Lithographic Facility (EBLF) which forms part of the SERC Central Microfabrication Facility at the Rutherford Appleton Laboratory. The advantage of electron beam mask-making is that several different ULA designs can be combined onto a single mask. With four student designs per ULA and up to five different ULAs on one mask it is clearly feasible to integrate an entire class of 20 students onto a single slice. The underlying principle is identical to that expounded by Conway[4], except that it produces semi-custom as opposed to full-custom ICs. If the scheme outlined above is successful, then the economics of using the ULA for teaching IC logic design begin to look very favourable indeed. At a conservative estimate the cost per student would probably work out to be approximately £20 to £25 for the optimum class size of 20. The possibility of using the ULA more extensively in research projects also presents itself.

ACKNOWLEDGEMENTS
The author would like to express his gratitude to the SERC for providing the funds and resources to support the M.Sc. course, and to Ferranti Electronics Ltd. for their collaboration in producing the ULAs. Thanks are also due to my colleagues in the Departments of Electrical Engineering and Electronics and Computation without whose encouragement and assistance this project would not have been possible. The invaluable work done by Mr. R. A. Cottrell, Computer Officer in the IC Design Laboratory, towards the setting-up and running of the laboratory is particularly worthy of recognition.

REFERENCES

[1] Morant, M. J., 'Low-cost silicon device fabrication in degree courses', *this volume*, p. 2.
[2] McKell, H. D., 'Production of microelectronic components at UMIST', *this volume*, p. 54.
[3] Mead, C. and Conway, L., *Introduction to VLSI Systems*, Addison-Wesley (1980).
[4] Conway, L., Bell, A. and Newell, M. E., 'MPC 79: A large-scale demonstration of a new way to create systems in silicon', *LAMBDA*, **1**, No. 2, pp. 10–19, (1980).
[5] Conway, L., 'The MPC Adventures: Experiences with the generation of VLSI design and implementation methodologies', *Xerox P.A.R.C. Technical Report VLSI*-81-2, (1981).
[6] Hicks, P. J., 'An introduction to semi-custom ICs', *Proc. of the First Int. Conf. on Semi-custom ICs* (London), (1981).
[7] Posa, J. G., 'Gate arrays — A special report', *Electronics*, **53**, No. 19, pp. 145–158, (Sept. 25th, 1980).
[8] Poole, K. F., 'A novel university approach to teaching microelectronics', *this volume*, p. 11.
[9] Murphy, B. T., Glinski, V. J., Gary, P. A. and Pederson, R. A., 'Collector-diffusion isolated integrated circuits', *Proc. IEEE*, **57**, pp. 1523–1527, (1969).
[10] Ferranti Electronics Ltd., '*Ferranti ULA — The design manual*', (1981).
[11] Mackintosh, I. M., 'New problems raised by logic arrays and their impact on the structure of the electronics industry', *Proc. First Conf. on Semi-Custom ICs* (London) (1981).
[12] Ramsey, F. R., 'Automation of design for Uncommitted Logic Arrays', *Proc. 17th Design Automation Conf.* (Minneapolis) pp. 100–107 (1980).

5

PRODUCTION OF MICROELECTRONIC COMPONENTS AT UMIST

H. D. McKELL
Department of Electrical Engineering and Electronics, University of Manchester Institute of Science and Technology, England

INTRODUCTION
The Solid-State Electronics Group in the Department of Electrical Engineering and Electronics at UMIST was established in 1962 and has always been concerned with the physics, manufacture and failure mechanisms of semiconductor devices. Apart from some research work on device production techniques in the early years, the very important area of manufacture has been mainly a teaching interest, supported from time to time by laboratory experiments and project work carried out in ordinary laboratory conditions, using standard laboratory equipment. For some years the Group has felt it desirable to establish a production facility in which the elements of the industrial process can be demonstrated, and in which a certain amount of more serious research work can be undertaken.

ESTABLISHMENT OF THE FACILITY
Some six years ago the Group was able to purchase, very cheaply, three triple-tube diffusion furnaces which were surplus to a manufacturer's requirements. These furnaces had been used to process 37.5 mm silicon wafers (the industry standard when the furnaces were originally bought) and had been out of use for some considerable time. Various schemes to bring these furnaces into use were considered, and rejected for one reason or another, usually lack of time or money or both. The situation in the summer of 1979 remained the same, the somewhat decrepit furnaces in store, and no clear prospect of their being brought into use. In June 1979, however, the Institute released a sum of money for the purchase of capital equipment. The Group submitted a bid for a number of clean air cabinets, and this bid was fortunately successful. The cabinets were ordered from Fell Clean Air Ltd. about Christmas 1979 and delivered in the spring of 1980. It is these cabinets which form the basis of the facility.

During the period June–December 1979, when the ordering of the clean cabinets was being negotiated, the department was invited to submit a proposal to the Science and Engineering Research Council (SERC) for a new M.Sc. course in Integrated Circuit Design. Discussions with members of the department, and with the Department of Computation made it clear that there was wide agreement that a considerable fraction of such a course should be devoted to device fabrication, and that this material should, if possible, include 'hands-on' experience of some of the processing involved. It was also decided that an important aspect of the course was to be the design of a system on a Ferranti U.L.A., which would be fabricated by Ferranti, but would be returned to UMIST for probing, dicing, mounting and bonding. These requirements had considerable influence on the equipment provided in the room. The UMIST M.Sc. course proposal was accepted, and the resultant SERC funding has made possible the purchase of suitable equipment.

THE FACILITY

The general layout of the clean cabinets is shown in Fig. 1. The room in which they are housed has been raised to a fair standard of general cleanliness by double glazing the windows, sealing the obvious gaps in the folding wooden partition which divides it from the laboratory next door on one side, and providing a welded P.V.C. floor. Two fans suck air into the room through filters. Each clean cabinet contains its own fan and filter unit, designed to blow filtered air downwards on to the working surface. It is this which provides the

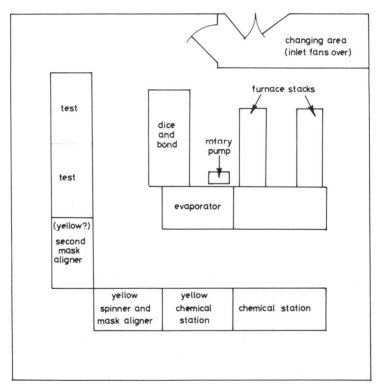

FIG. 1   Approximate layout of clean room (scale 0.25 inch to 1 foot).

clean working conditions. The cabinets designated as chemical work stations are provided with fume extraction through a perforated polypropylene work surface. The extraction duct is polypropylene throughout because of the corrosive nature of some of the products (HF) being exhausted. The relative speeds of the fume extraction fan and the air inlet fans are chosen so that the whole room is maintained at a small positive pressure which, we believe, will keep out the dirty outside air. Deionised water is provided on the chemical work benches from an Elga deionising column.

The sketch shows the position of two three-tube furnaces, and the current allocation of clean cabinet space to other activities. The furnaces have been largely rebuilt, and all the control wiring and thermocouple connections inside the cabinets have been replaced, as the insulation on the existing wiring was all more or less useless. We have replaced the ceramic liners inside the furnace elements with liners of sufficient internal diameter to permit the use of 50 mm silicon wafers. As yet, we have not bought 50 mm quartzware, and have continued to use our existing stock of 25 or 37.5 mm wafers, or sections cut from 50 mm wafers. We had some difficulty in deciding on 50 mm as a new standard size rather than 75 mm, arguing that 50 mm was now so old-fashioned in industrial terms that its supply might quite soon become a problem. However, to modify the furnaces to accept 75 mm slices would have entailed fitting new elements, and this seemed to be an unjustified expense. This, however, remains an option for the future. The third furnace stack has been used as a source of spares.

Wafers are loaded into and removed from the furnaces from within the clean cabinet (Fig. 2). Furnace gas is piped into the other end of the tubes, currently from standard cylinders stored in the room next door. It is hoped that in the very near future this arrangement will be improved by the provision of a liquid nitrogen dewar to act as a gas source. This will allow the furnaces to be run, as they should be, with a continuous standby flow of nitrogen, allowing the

FIG. 2 Loading wafers into an oxidation tube.

exclusion of dirt. We have consciously decided not to buy and install our new 50 mm quartzware until we are assured of a continuous nitrogen supply. All diffusions are carried out using Emulsitone spin-on sources. While this is a somewhat inflexible system, it means that there are no dangerous gas or liquid sources to handle and no problems with poisonous exhaust gases.

A standard oil-pumped evaporator has been built up by the department's engineering workshop staff and installed in a clean room table. A 12-inch diameter bell jar is pumped by a 4-inch diameter Edwards diffusion pump backed by a 150 l/min rotary pump. The rotary pump is positioned as shown, outside the clean cabinet. The system pumps from atmospheric pressure down to a pressure adequate for the evaporation of aluminium or gold in less than 5 minutes. Two simple resistively-heated evaporation filaments are provided.

The two six-foot clean cabinets marked 'yellow' are intended for photo-resist processing, and are provided with yellow fluorescent lighting and yellow perspex end panels. It has unfortunately not proved possible to work consistently with photo-resist in these cabinets, as it seems that the spectrum emitted by the yellow tubes contains an unacceptable amount of ultra-violet. It is believed that this problem is on the verge of solution with the recent acquisition, (unfortunately not yet fitted) of some Kodagraph filter material. A Headway twin-head resist spinner was bought new and is used for spinning on both resist and diffusion sources. The preferred resist is Shipley AZ 1350H.

The Group has two mask aligners, a Cobilt model CA-800 bought second-hand (Fig. 3), and a very old machine made 'in-house' by Ferranti and given to the Group some time ago. Currently this is the only aligner actually working, although the Cobilt awaits only its establishment in the clean cabinet and connection to the supplies, particularly of compressed air. It seems almost certain that we shall have to continue to provide both mask aligners, and this will imply the conversion of a further six-foot clean cabinet to yellow lighting.

Wafer probing is carried out using a manual system bought from Research

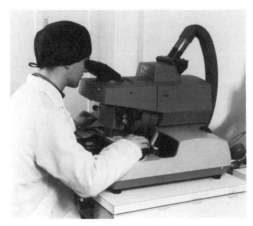

FIG. 3   Mask aligner for 50 mm and 75 mm wafers.

FIG. 4   Multiprober in use.

Instruments, fitted with twelve probes and an inker (Fig. 4). The manual location of each chip to be tested under the probe array is not unbearably wearisome given the relatively small numbers we are required to test.

Wafers are diced by diamond scribing and breaking. An automatic diamond scriber has recently been obtained, which has made the successful scribing of 3-inch U.L.A. wafers very easy. The previous manual scriber was almost impossibly difficult to align with the narrow (~ 80 μm) scribe channels.

Dice have been bonded into 40-pin ceramic dual-in-line packages using silver loaded epoxy (Abelbond 36-2). We have bought an automatic dispenser for the epoxy, but this has, as yet, not proved very reliable in use.

Dice are bonded to the package connections using a Kulicke & Soffa Model 479-3 Thermosonic bonder. The actual positioning and formation of the bonds with this machine has been virtually trouble-free, but ball formation has been a great deal less reliable than one would have expected. However, the equipment does allow us to probe-test, dice, mount and bond the Ferranti U.L.A.s designed by our M.Sc. students[1] reasonably quickly and quite successfully.

TEACHING ACTIVITIES

The clean room has been used during the past year by students on the second and third years of the undergraduate course, and by M.Sc. students on the Solid-State Electronics and Integrated Circuit Systems Design courses. A variety of different device structures have been fabricated, simple gold-on-silicon Schottky diodes, diffused junction diodes, MOS capacitors, and simple solar cells. As yet, the processing is not fully under control, and the results are somewhat variable. In the case of the Schottky diode manufacture, which has been carried out much more often than the other processes (it is a second-year experiment) it seems that over the last month or so the devices have been consistently closer to the theory than those manufactured in previous years in less clean conditions.

FUTURE TEACHING ACTIVITIES

We have no intention of setting up a line to manufacture integrated circuits of any great degree of complexity. We do not seem likely ever to have the manpower to do this, even if it were desirable. We intend to work towards the routine production of small scale integrated circuits of the complexity of, say, a single flip-flop. Initially the chosen technology will be metal gate p-MOS with an ultimate aim of polysilicon gate n-MOS. Minimum feature sizes will probably never be less than 5 µm.

In connection with the U.L.A. design activity[1] there are plans to take over the definition of the metal pattern. The existing equipment should be adequate for this purpose.

RESEARCH

The clean room facilities are widely used to support the research work of the Group, mainly in the production of more or less simple structures for use as experimental specimens. This often consists of no more than making a single rectifying or ohmic contact. Many of the samples handled in this way are small, and of irregular shape. It turns out that the old-fashioned mask aligner is much better suited to handling this kind of sample than the newer one, designed specifically for 50 and 75 mm wafers. It is to support this kind of activity that the old mask aligner is being kept in service.

ACCESS TO THE CLEAN AREA

Because the facility is aimed at providing hands-on experience, that is, letting students actually align their own patterns, bond their own chips etc., it is not possible to restrict access to a limited number of clean room personnel, nor to insist on a very high standard of clean-room clothing. Overshoes, hats, and nylon laboratory coats are provided, and laundered on a regular basis. If this proves not to produce an adequate standard of cleanliness then these practices will have to be reviewed.

REFERENCE

[1] Hicks, P. J., 'A practical approach to digital integrated circuit design using uncommitted logic arrays', *this volume*, p. 38.

6

MICROELECTRONICS TEACHING AT THE UNIVERSITY OF CANTERBURY, NEW ZEALAND

L. N. M. EDWARD
Department of Electrical Engineering, University of Canterbury, Christchurch, New Zealand

1 INTRODUCTION

This paper is written with two aims: firstly to reveal details of microelectronics activity at this university; and secondly, by describing the laboratory in some detail, to encourage others who may be wishing to attempt something similar.

Early attempts to establish a microelectronics facility in 1973 were rewarded with generous verbal support from both government and industrial bodies but no finance[1,2]. By mid-1975 the author decided on a go-it-alone, build-it-yourself solution and a written appeal to individual firms yielded $1300 to purchase a DEK 65 screen-printer, whilst grants have been received from several firms for prototype developments.

One object is to help to introduce microelectronic technology to New Zealand (NZ) industry and therefore the product must be commercially viable, which requires professional-quality equipment.

Prototype runs of 100 or more units have been successfully produced.

The first thick-film circuit was fired on 14 November 1978 and the first microelectronics course commenced on 24 May 1979 with 6 final-year students and one postgraduate. Table 1 shows the growth in numbers since then.

To date, only thick-film hybrid microelectronics has been taught, but we intend to develop a capability in very large scale integrated circuit (VLSI) engineering once the necessary design software has been established.

We believe that thick-film experience demonstrates to students most emphatically both the versatility and potential of microelectronic technology, while the laboratory is also justified by the service provided to university researchers, public bodies and industry.

TABLE 1 *Student numbers.*

Year	Undergraduate	Postgraduate	Number of circuits
1979	6	1	7
1980	7	4	11
1981	26	6	7

Other thick-film facilities in N.Z. are at the Department of Scientific and Industrial Research (D.S.I.R.) Physics and Engineering Laboratory in Lower Hutt, Electronic Microcircuits Ltd. in New Plymouth, and Philips Electrical Industries N.Z. Ltd. in Wellington.

2 COURSE STRUCTURE

Today's rapid pace of technological change is placing severe stress on our course structure because of the need for continuity between successive academic years as the student progresses towards his degree which involves one intermediate and three professional years. Conventional first-degree courses are unable to include new material without rejecting something essential.

Historically, the University of Canterbury Batchelor of Engineering (BE) degree has been broadly based with little specialisation, because most graduates are employed by one of the major government departments such as N.Z. Electricity, N.Z. Post Office, N.Z. Broadcasting Corporation, N.Z. Railways, Ministry of Transport, Ministry of Aviation and others.

The first degree cannot carry a topic such as VLSI engineering to the point of actual design without compromising the needs of major employers, whilst limitations of staff, facilities and time-tabling currently rule out parallel degrees in (say) Power, Control, Electronics, Communications, etc. A flexible degree structure based much more on individual student preferences would require course material to be presented in smaller, self-contained units. Such a structure would be resisted by many, in the belief that N.Z. still requires the majority of its engineers to have a broadly-based initial training which will better prepare them to interface with the many loosely-related disciplines. Therefore we believe that a Masters degree course is the most appropriate place for such topics.

2.1 *Hybrid microelectronics course*

The present course involves:
(a) Introduction and overview of discrete, thick-and-thin film, and monolithic technologies.
(b) Components, their characteristics and limitations.
(c) Design philosophy and procedures for initial sizing; layout of resistors, capacitors and inductors; temperature effects; crossovers and multilayers; parasitic effects.
(d) Thermal design and cooling; add-on components; chip and wire-bonding, reflow soldering; epoxies and their uses.
(e) Testing; rework; encapsulation and packaging, environmental effects.
(f) Artwork generation; photo-fabrication; screenmaking.
(g) Screenprinting, conductor, resistor and dielectric paste characteristics.
(h) Drying and firing; trimming of resistors and capacitors; drift and aging; reliability.
(i) Computer-aided-circuit design, simulation and optimisation using LINSIM[8].

(j) Production-engineering considerations.

Each student receives, on loan, a 76-page Laboratory User's Manual describing the processes and components available, and a 50-page data appendix.

The course, open to Masters students, occupies the first 10 lectures of a 48-lecture 3rd Professional year *Electronics and Instrumentation* module. During the first term each student designs and prepares the artwork for a circuit to fit within a 1-inch square substrate, and in the second term the best of the designs are fabricated and tested. In 1979 and 1980 the classes were small enough for each student to fabricate his own circuit, but in 1981 numbers were so overwhelming that a competitive design approach evolved. The Masters students have more flexible work schedules and all fabricate their own circuits.

3 REPRESENTATIVE PROJECTS

3.1 *Undergraduate and postgraduate*
Polyphase audio phase-shift network for S.S.B. generation
Digital interface to tape-recorder
Search and lock transceiver scanner
Miniature A.M. radio using a ZN414 integrated circuit
Proportional temperature controller
Special A.G.C. amplifier for a blind-aid
Exponential (antilog) voltage-to-voltage converter
Switched precision voltmeter preamplifier
Ultrasonic preamplifier for a blind-aid
Electronic dice
Dual high-speed Schmitt trigger
DC to DC inverter

The above projects were constructed in 1979 and 1980 when each circuit was proposed by the student who, because he was allowed to keep it, was strongly motivated to succeed. All circuits worked and most were fully operational.

In 1981 each student designed a biquad notch filter to work between 2 and 10 kHz. Circuit layout and performance became competitive, with LINSIM[8] heavily involved in circuit optimisation where the influences of operational amplifier bandwidths and the capacitive effects of crossovers and proximate conductors were dramatically seen. Notch depths of 20 dB and Q-factors of 15 to 20 were obtained when resistors were pre-trimmed air-abrasively to suit measured ceramic chip capacitor values. No active trimming was performed.

The relative effects, on chip-capacitor 7XR dielectrics, of conductive epoxy bonding and reflow-soldering highlighted, as was intended, both the temperature-coefficient and the longer-term shift-and-drift characteristics. The need to choose very carefully filter components and the capacitance/volume/stability tradeoffs were clearly demonstrated.

3.2 Industrial

An important, but frequently unrecognised, facet of education is assistance to industry. In this wider industrial sense a successful commercial involvement becomes known in the most unexpected quarters.

For example, animal capture, particularly red deer and wapiti (elk), by means of a drug dart fired from a helicopter, hitherto required that the animal be kept in view for up to 15 minutes until the drug took effect. A thick-film VHF transmitter beacon (pulsed) in the tail of the dart now allows the animal to be located quickly in the thickest bush, so the animal can be left to go quietly to sleep instead of being terrorised by the continued presence of a noisy helicopter. Both capture-and survival-rates have dramatically improved through this innovation.

The transmitter is cylindrical, 0.410 inch in diameter and 1.150 inch long including crystal and battery. It will withstand 20,000 G dart impact and impressed our students greatly!

4 THE LABORATORY AND MAJOR PROCESSES

Figs. 1 and 2 show the floor-plan and a view of the furnace room.

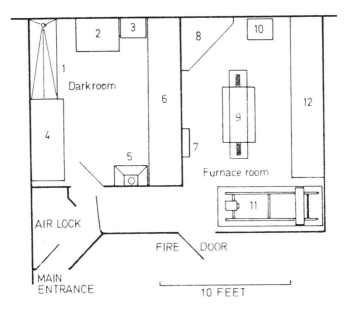

FIG. 1 Microelectronics laboratory layout. (1) Wet bench; (2) Clean (laminar-flow); (3) Refrigerator; (4) Work bench (viscosity, light-section microscope); (5) Fume cupboard; (6) Work bench (screen-printing, drying, assembly); (7) Furnace air control and purification; (8) Ultrasonic wire-bonder; (9) 5-Zone belt-furnace; (10) Electronics rack (general laboratory systems, furnace-control, etc.); (11) Photo-reduction system; (12) Work bench (assembly, testing, etc.).

FIG. 2   Furnace room.

4.1   Getting started

From the outset I decided to design and have built in our workshop as much equipment as possible to fully professional standards. The available cash was spent on materials and the design/labour costs were hidden in salaries. From mid-1975 it took three and a half years to reach the stage of firing the first trial circuit, but results since then have fully justified the initial decision to do it properly the first time.

I will now briefly describe some of these 'home-built' units.

4.1.1   *Furnace*   This comprises a 3-inch diameter, 40 inches long tubular silica muffle threaded by a 2.5 inch wide continuous belt which carries the product steadily through five heated zones. Each zone is controlled by one of five PID temperature control loops implemented by a multi-tasking 8085 microprocessor. Temperature control is to $\pm 0.5°C$ and the zones blend to form a highly stable and reproducible time/temperature firing profile with peak temperature up to $1100°C$, though most firing is at $850°C$ peak. The belt is made from 1000 helical sections of nichrome V wire and driven at precise speed by a steppermotor. Firing atmosphere is dried and purified air at about $-50°C$ dewpoint and waste furnace gases are extracted at the product-input end by an air-driven venturi. Air curtains prevent laboratory air entering or furnace gases polluting the laboratory. Provision has been made for firing in a nitrogen atmosphere.

4.1.2.   *Photo-reduction of master artwork*   A Sinar-f monorail camera with a Rhodenstock SIRONAR 150 mm lens is mounted on a rigid bed which carries a travelling lightbox, the whole being mounted on rubber shock-mounts. Rubylith cut-and-strip artwork masters up to 22×32 inches can be photographed on 5 × 4 inch cut film at from 11:1 reduction to 1:1.2 enlargement. Final mask resolution is adequate for thin-film circuits with 0.001 inch wide conductor tracks.

4.1.3 *Screens and exposure*   Using a Philips HPR-125 mercury-vapour ultra-violet lamp, about 15 minutes exposure is required for Autoline TF emulsion on 330-mesh orange polyester screen material.

Our working standard is 0.010 inch conductor tracks and spaces but 0.005 inch dimensions have been successfully used. Screen material is stretched for several weeks on EL-303 CAM-LOC* frames and then epoxy bonded to DEK aluminium screen-frames. Tension is closely monitored and, with careful cleaning, we are re-using the same screen with up to 14 different circuits before renewing the mesh.

4.1.4 *Substrate drying*   A 6-inch wide continuous canvas belt driven by a small stepper-motor carries the freshly printed substrates through a covered 'levelling' section and then under an infra-red sunlamp for about 15 minutes at 125°C, the process all being purged by filtered air. This same equipment serves to quick-dry substrates after in-process rinsing in distilled water during later assembly of discrete components etc.

4.1.5 *Laminar-flow workstations*   A 5-inch water-gauge, 250 c.f.m. air blower supplies two laminar-flow workstations within the laboratory; one contains the ultrasonic wire-bonder, the other a general-purpose stereo-zoom microscope workstation. A third workstation in the test area outside the laboratory has its own blowers built in. Thus assembly and test of unprotected circuits takes place in about Class 1000 conditions.

4.1.6 *Reflow soldering and dip tinning*   The reflow hotplate is a 3-inch diameter, 2-inch long, aluminium cylinder which was first dipped twice in Sauereisen No. 78 cement, wound with a nichrome V element and redipped several times. A Type K thermocouple imbedded just beneath the top surface controls temperature to $225 \pm 1°C$.

Dip soldering and tinning are done in a small solderpot made as follows. A 250 ml Pyrex beaker is set into a $1\frac{5}{8}$ inch deep cavity in an aluminium cylinder using Dow-Corning RTV 3120 silicone compound (red). Otherwise, construction and temperature control are as for the hotplate. We use (percentages) 62 tin, 36 lead, 2 silver solder alloy at 230°C, with excellent results on silver-palladium conductors which have been coated with mild resin flux. Post-reflow cleaning in FREON** TWD-602 with a few seconds ultrasonic agitation followed by a short ultrasonic scrub and vapour-rinse in FREON TMS produces ultra-clean assemblies ready for encapsulation[7].

---

*Registered trade mark of the Advance Process Supply Co.
**Registered trade mark of Du Pont de Nemours and Co. (Inc.).

4.2 Equipment purchased

4.2.1 Ultrasonic wirebonder

Our Kulicke and Soffa Model 484EE bonder uses 0.001 or 0.0015 inch diameter aluminium/1 percent silicon wire with a 60° concave-tip bonding tool. Such bonders are sensitive to vertical acceleration caused by floor vibrations. As little as 0.1 G may cause up to ± 100 percent fluctuation of static bonding tool pressure on the wire at the instant of bonding, causing poor bonds. Therefore our bonder has been placed on a 500 lb laminated-steel (16 SWG sheets) table carried on three pneumatic supports, each with an equivalent spring-rate of 14 inches.lb.f^{-1}, which gives the whole a natural vertical resonance of about 0.4 Hz. Each support is independently maintained at proper height by spool-valve air-servomechanisms which correct for variations in operator arm weight. Wirebond pull-strength standard-deviation is typically less than 20 percent about a 12 grams average[4].

4.2.2 Resistor and capacitor trimming

An S.S. White Model K Airbrasive unit together with a dust collector and small cyclone cleaner (home-built) and a manually-controlled X–Y table are assembled on a 4 ft × 2 ft 6 in × 36 in high trolley. The substrate holder has 0.6 inch of pneumatically powered vertical movement for loading and unloading. Resistors 0.05 inch or more in width are easily trimmed to 0.1 percent or better.

Ceramic chip capacitors are trimmed by abrading a crater in the center.

4.2.3 Instrumentation and measurement

Quality control is an important part of microelectronic engineering education, so we monitor and adjust for solvent loss from thick-film pastes using a Brookfield RVT Viscometer.

During the past three years our resistor-calibration substrates (Fig. 3) have consistently yielded standard deviations of 3 to 4 percent for 1Ω to $100k\Omega$ per square pastes and less than 10 percent for $1\ M\Omega$ to $10\ M\Omega$ pastes.

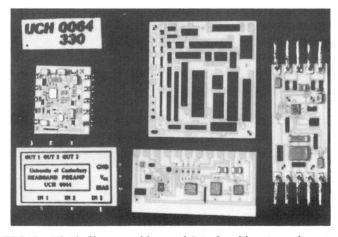

FIG. 3   Thick-film assemblies and 1-inch calibration substrate.

The most useful parameter for on-line screen-printing control is wet print thickness which must be measured using non-contact methods[3]. We use a Zeiss light-section microscope to obtain ± 0.5 micrometer accuracy in the 50 micrometer thickness range.

Screen tension measurement is made with a 1 lb weight loading a dial-gauge which presses a 0.5 inch radius foot against the exact screen centre. Tension is read as a deflection on the gauge which has had its return spring removed so that, without the weight, the foot rests very lightly against the screen causing a fixed offset of only 0.002 inch. We find 0.1 inch deflection suitable and renew the polyester mesh when, for a 6×8 inch screen, deflection exceeds about 0.12 inch.

Screen deflection and snapoff are set at the printer using a similar gauge with a foot simulating the squeegee edge and this permits a predetermined correction to the snapoff to compensate for various screen tensions[5].

4.2.4 *Screen-printing registration* A very practical method of registering a print is to place over the substrate a small (2×2 inch) square of clear mylar sheet (0.002 inch thick) which earlier was peeled from the Autoline direct-indirect emulsion at screen-making time. Controlled vacuum leakage of the substrate holder firmly secures the sheet and a trial print is made onto it. Using a torch-magnifier ($10 \times$), registration errors between the print and underlying substrate down to 0.001 inch can be noted and the necessary printer adjustments made. Each test print is wiped clear of the mylar using lint-free tissue, excepting the last, which is dried and mounted in the laboratory notebook as a convenient position reference map referring to places where various wet, dry and fired print thickness measurements were made.

4.2.5 *Thick-film pastes* The substantial investment of about $1200 in resistor, conductor and dielectric pastes together with the un-knowable cross-compatibility between various manufacturers obliged us to use a single source which happens to be Electro-Science Labs, Inc. (ESL).

All pastes and photographic film are stored at $-4°C$ in a domestic refrigerator, whilst epoxies are kept in the freezer unit. Resistor pastes up to 1 MΩ per square, dated March 1979, are still in use despite the nominal life being given as 9 months. Conductor and dielectric pastes seem to keep indefinitely in the refrigerator.

4.3 *Resistor paste calibration*[3, 6]

We calibrate in two aspect-ratio ranges: 1 to 10 and 0.1 to 1 using the model of Fig. 4 where $R_c/2$ is the so-called contact resistance and R the apparent bulk sheet resistance. The total resistance R_t is then

$$R_t = R_c/2 + R + R_c/2 = R_c + R_s N$$

where R_s is the apparent sheet resistivity and N the number of squares between the resistor terminations.

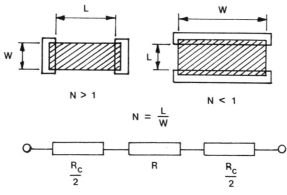

FIG. 4   *Resistor model.*

Test resistors in the range $N = 1$ to 10 have constant W for varying L, whilst for $N = 0.1$ to 1 L is constant and W is varied.

N takes values of 0.1, 0.2, 0.4, 0.6, 1, 2, 4, 6, 10 making up 39 different resistor sizes per 1-inch square substrate (Fig. 3). The smallest resistor is 0.01×0.01 inch and the largest 0.4×0.1 inch, N being restricted for larger resistors.

A computer-controlled digital multimeter is used to measure each test resistor and then the design parameters R_c, R_s are computed for each $W(N > 1)$ or $L(N < 1)$ together with histograms and temperature-coefficients, from which are produced data sheets for the particular paste lots. Recalibration is at maximum intervals of about three months.

As-fired resistor mean values are typically within 10 percent of target and ratios within 3 percent. When the present hand-operated printer is modified by mechanising the printing stroke, it may be worthwhile adopting the model of Zarnow[3] for better as-fired accuracy.

4.4 *Add-on components*
A wide range of bare silicon dice is available, which we attach to the substrate using either silver-loaded conducting, or non-conducting, epoxies. No eutectic bonding is practised because of the prohibitive cost of gold conductor pastes and this is a restriction on the power handling capacity of bonded power transistor dice.

4.5 *Packaging*
Metal-and-glass hermetic packages are expensive, difficult for us to assemble and test, and can be, themselves, sources of moisture and contamination.

Potting in general-purpose plastics or epoxies may cause thermal stresses which fracture or tear away wirebonds and add-on components and it is not hermetic.

After a long and erratic search, we are currently using HYSOL* EO-1017

---

*Registered trade mark of HYSOL DIVISION, The Dexter Corp'n.

thixotropic one-part epoxy/silicone compound directly over wirebonds and all. For example, a substrate carrying pins in DIL configuration can be hand-coated, using a syringe, to form a slightly convex surface onto which is placed a matching cover of alumina substrate material carrying appropriate identification printed and fired using time-expired resistor paste. After 16 hours curing at 150°C there emerges a quite professional-looking ceramic DIL package which seems to be at least as good environmentally as a commercial plastic DIL package (Fig. 3).

5 OVERALL PLANT COSTS

In round figures, at historic cost between 1975 and 1978, we expended about $21,250 (N.Z. dollars) on new equipment. Raw materials to the value of about $4,500 were fabricated by our technicians into equipment valued at $18,650. When structural changes to the laboratory and equipment donated are added in, the total comes to about $50,000.

6 FUTURE MICROELECTRONICS DEVELOPMENTS

The laboratory, though at present supporting thick-film technology, was conceived more widely. A thin-film capability will soon be commissioned, its workload to be shared between the optics and microelectronics laboratories.

The current trend towards custom VLSI fabricated at silicon foundries and cooperation in multi-project-chip programs with the Commonwealth Scientific and Industrial Research Organisation of Australia (CSIRO) removes the need for silicon fabrication here at Canterbury. We expect to introduce VLSI studies in the near future, once the essential computer-aided-design (C.A.D.) tools are established.

7 CONCLUSION

Our laboratory is a well-established and reliable facility serving many kinds of user.

As teaching support it brings an essential realism to the discipline of microelectronics at a level readily comprehensible to the beginning engineer at a time when an enormous amount of manual design skill and effort is being entombed in C.A.D. software packages; skills which will probably never become known to students now graduating.

Most of our students eventually acquire some skill and confidence as they see that neither luck nor witchcraft but rather careful application and common sense are the determinants of success. Our N.Z. students, thus encouraged, will realise that no technology created by man should be beyond their compass and will have taken an essential first step towards parity with the developed nations; towards the immense system design capability and freedom which is coming within reach as the new VLSI methodology and design tools become available.

REFERENCES

[1] Edward, L. N. M. and Byers, D. J., 'Hybrid microelectronics in New Zealand,' *New Zealand Electronics Review*, 7, pp. 13–14 (September 1973).
[2] Edward, L. N. M. 'New microelectronics laboratory', *New Zealand Electronics Review*, **12**, pp. 43–47 (August 1979).
[3] Zarnow, D. F., 'A new approach to thick film resistors', *ISHM Proceedings of the 1979 International Microelectronics Symposium*, pp. 32–39 (November 13–15).
[4] Cheriff, F. J. and Salzer T. E. 'High reliability ultrasonic wire bonding,' *ISHM International Microelectronic Symposium Proceedings*, pp. 221–27 Sheraton-Boston Hotel, (October 21–23, 1974).
[5] Ottaviano, A. V., 'Repeatability in screen printing hybrid microcircuits', *ISHM 1969 Hybrid Microelectronics Symposium Proceedings*, pp. 253–62 (September, 1969).
[6] Kuo, C. Y., 'The contact resistance in thick film resistors,' *ISHM 1969 Hybrid Microelectronics Symposium Proceedings*, pp. 263–69 (September 1969).
[7] Creter, P. G. and Peters, D. E., 'A new method for cleaning microelectronic substrates', *ISHM Proceedings of the 1977 International Microelectronics Symposium*, Baltimore Hilton and Civic Centre, pp. 281–86 (October, 1977).
[8] Edward, L. N. M., 'LINSIM, A linear electrical network simulation and optimisation program', *this volume*, p. 138.

7

LINEAR INTEGRATED CIRCUIT DESIGN IN THE CURRICULUM

H. E. HANRAHAN and S. J. WEST
Department of Electrical Engineering, University of the Witwatersrand, Johannesburg, South Africa

INTRODUCTION
The design of very large scale integrated digital circuits has advanced very rapidly in recent years. This has been aided by the fact that most parts of a given digital system can now be implemented in a single technology, and that logic and memory can be mixed in the same integrated circuits. A second important factor in this development has been the availability of powerful design automation tools. Very large scale digital integrated circuits are made up of extremely large numbers of individual devices and are designed to such tight geometrical constraints that human designers cannot manage the design process without the aid of the computer. Techniques have been evolved for working at higher levels of system architecture in a highly automated design environment[1]. The concept of the silicon foundry which accepts integrated circuit designs in a specified format and merely performs the manufacturing steps, has become well established. This decentralization of integrated circuit design has made it possible for students at a number of universities to become involved in the design of VLSI integrated circuits, to date, largely digital in nature[2].

The tendency of analogue circuit design to lag behind that of digital circuits was pointed out by Gilbert[3]. However, he also stresses the fact that because analog circuit design is not limited in its capability, and the demands that will be made on it, other than by natural limitations such as noise, there will continue to be a demand for newer, more precise analogue integrated circuit designs. Recent reviews[4,5,6] reflect the continued importance of analogue and mixed analogue and digital integrated circuits in the fields of communications, electronic components and semiconductor device technology. The opinion has been expressed[7], that analogue integrated circuit designers could soon become an 'endangered species' at a time when substantial developments in sophisticated analogue and digital microsystems is taking place.

It is clear that there will be substantial demand for linear integrated circuit design in a number of fields:
(a) High frequency linear applications
(b) More precise versions of conventional linear integrated circuits
(c) Communications
(d) Signal processing

(e) Instrumentation
(f) Consumer and automotive electronics

The recent experience in the participation of universities in silicon foundry projects and the production of complicated digital integrated circuits to students' designs prompts a re-examination of the position of linear integrated circuit design in the curriculum. This paper considers the issues involved, the tools that are required and draws comparisons and highlights differences between a digital and analogue integrated circuit design.

LINEAR AND DIGITAL INTEGRATED CIRCUIT DESIGN

At the device level, the linear integrated circuit designer is faced with devices which have relatively complex electrical characteristics which exhibit substantial temperature dependence. The devices operate in the continuous rather than switching mode, and the non-linearity and temperature dependence of their characteristics is invariably a significant consideration in design at this level. Devices for digital integrated circuits, on the other hand, operate in a binary fashion and allow the user to move to the next level of circuit complexity with the device characteristics taken care of by means of a relatively small number of parameters such as loading rules, propagation delay, trigger voltage and definition of logical 1 and 0 states.

Just as the gate and bistable circuit represent the second level in a digital integrated circuit, in the analogue world combinations of basic devices such as current mirrors, differential pairs, cascode connections, give the analogue circuit designer a repertoire of basic building blocks for more complex circuits. Such building blocks are significantly more complex to characterise than the corresponding gate level of a digital circuit.

The next level of complexity from which a comparison can be made, is at the small to medium scale integrated digital circuit, and the level of complexity involved in a circuit such as an operational amplifier. The number of external specifications, particularly those that have to be given in the form of a graph rather than a spot value, for an operational amplifier is substantially greater than that for a digital circuit such as a decade counter, having approximately the same total number of transistors. In the latter, the temperature dependance is accounted for by specifying an overall range in which the device will work satisfactorily, subject to the observance of loading rules. This contrasts with the temperature-dependent performance of a linear circuit which gives a continuous variation in the performance of the circuit with temperature and signal level.

Moving to higher levels of integration, the problem with digital circuits lies mainly in handling the complexity of the circuit and translating the circuit design into manufacturing information. Recent advances in design automation have provided a relatively effective solution to this problem. While the scale of integration of analogue circuits, as measured by the number of devices per chip, has not increased as fast as that of digital VLSI, it is an area where advances in design are necessary to ensure progress.

ISSUES INVOLVED IN LINEAR INTEGRATED CIRCUIT DESIGN

At the device level, the modelling of devices has advanced sufficiently far to enable relatively complex collections of devices, e.g. bipolar transistors, to be studied via circuit analysis packages such as Spice2[8].

This circuit analysis program supports the definition of subcircuits. That is, where a configuration of devices is used often, a subcircuit can be defined and 'called up' at a block level. The analysis of the analog circuits dictates that the subcircuit cannot be replaced by an equivalent model, as this would be too complicated in the general case. The circuit has, instead, to be analysed at the device level. That is, the subcircuit must be replaced, for analysis, by the devices it is made up of[9].

The designer of analog circuits must have a good feel for the function, operation and limitations of any subcircuits he is using. To achieve this, he requires a good fundamental understanding of the devices he employs. Special approaches and techniques for building blocks have emerged. In order to give the designer building blocks which overcome the non-linearity of the device, and provide a great degree of temperature compensation, a class of such circuits is the translinear circuit[10]. This example illustrates the need for detailed knowledge at the sub-building block level on the part of the designer, as it may be necessary to have a detailed continuous model of each port of the building block.

An important issue in linear integrated circuit design is the question of specification. Relative to digital VLSI, the specification of analogue integrated circuits is quite complex and many pitfalls exist. A common example is the analogue to digital converter which may have an impressive specification in terms of the number of bits to which it resolves, but few users realize that it may not have a monotonicity specification to match. A thorough specification of analogue integrated circuits requires an intimate knowledge, down to the device level. The simple reason for this is that, at any pin to the integrated circuit package, a direct connection may be made to a device which operates in a linear mode, and the characterization of a port consisting, for example, of the base of a transistor and a common terminal, is relatively complex and needs several parameters to describe it. In digital integrated circuits, all these detailed considerations are absorbed into logic level conventions and loading rules. In the case of ratio design of digital circuits the number of 'analog' specifications are reduced. The designer is interested in '1' level, '0' level, and trigger voltage, so that noise margins may be evaluated. The designer must also concern himself with rise and fall times which are a function of the capacitive loading of each gate and the gates output current capability. A further complication in the specification of linear integrated circuits is that, under certain conditions, non-linear behaviour can be exhibited — for example when an operational amplifier is operating in a slew-rate limited mode.

The availability of circuit design aids, such as Spice2, is of vital importance in aiding the designer to explore the performance of his circuit. It must be made available on an interactive basis, giving the designer a short turnaround.

To make it possible for students to tackle all but the most simple linear integrated circuit designs, it is necessary to have design automation for laying out the components and interconnections and for the production of manufacturing information for submission to the analogue silicon foundry.

To build the confidence of a circuit designer, practical testing of circuits using breadboard components is very instructive. The student in circuit design can learn very efficiently from hands-on experience at the device level. As circuits become more complicated, however, breadboard tests become more difficult. Stray capacitance, interference, and poor temperature tracking, reduce the accuracy of breadboard tests. For these reasons, computer simulation in many cases gives a more useable and a more accurate indication of circuit performance. Breadboarding is inaccurate because it is impossible to duplicate an integrated circuit using breadboard components[8].

The silicon foundry concept and the availability of substantial design aids for digital integrated circuits have resulted in a situation where the custom design is, in many cases, preferable to the use of uncommitted arrays of logic elements. It is not clear when the same situation will arise in the case of linear integrated circuits, and the silicon foundry can probably service the majority of students' needs by means of uncommitted analogue integrated circuit[11,12,13,14,15].

The question of the testing of the prototypes produced by the silicon foundry is of paramount importance[16]. The principal difficulties are the very high cost of automatic test equipment for analogue integrated circuits and the fact that most available systems are geared to a production environment rather than to circuit development. There is, clearly, a need for highly-flexible programmable automatic test equipment to support this activity.

To feed results efficiently back to the designer, a large number of tests will need to be repeated at different operating points and under different operating conditions. The tests may also be repeated on a number of samples of the prototype circuit.

The automation of tests, together with processing of results to present them in a meaningful format, is attractive. A small analogue circuit test system might include programmable signal sources and measurement equipment, together with a microcomputer as a system controller. In this way, results from SPICE2 simulations and measurements on prototype integrated circuits can be compared.

CONCLUSIONS

The continued importance of analogue integrated circuit design, together with the apparent lack of persons interested in this field, provides a serious challenge to universities. Examples of limited activities in this field are cited in the literature[13,14,15]. Smith[17] argues that a systematic approach to the design of complex circuits from the component level is needed. A possible approach could be adapted from the methodology of computer science and software engineering. The steps would be the development of a descriptive language enabling the designer to progress from the discrete component through various levels of architecture, to a description of the fully integrated and encapsulated functional integrated circuit. This has been achieved, to a large extent, in the digital area, but the challenge remains in the field of linear integrated circuits.

REFERENCES

[1] Mead, C. A. and Conway, L., *Introduction to VLSI Systems*, Addison Wesley (1980).
[2] Marshall, M., Waller, L. and Wolff, H., 'The 1981 Achievement Award', *Electronics*, 102–105 (20 October 1981).
[3] Sheingold, D. H., (ed)., *Non-Linear Circuits Handbook*, Analog Devices Inc., Norwood Mars (1976).
[4] Hindin, H. J., 'Technology update: Communications', *Electronics*, 216–225 (October 20, 1981).
[5] Beresford, R., 'Technology update: Components', *Electronics*, 144–147 (October 20, 1981).
[6] Posa, J. G., 'Technology update: Semiconductors', *Electronics*, 116–123 (October 20, 1981).
[7] Mokhoff, N., 'Analog IC designers could soon become an endangered species', *The Institute*, p. 4 (April 1981).
[8] Chua, L. O. and Lin, P. M., *Computer Aided Analysis of Electronic Circuits*, Prentice Hall (1975).
[9] Nagel, L. W. and Pederson, B. O., SPICE2 (Simulation Program with Integrated Circuit Emphasis) *University of California, Berkley Electronics Research Laboratory Report MEM ERL M520* (1975).
[10] Seevinck, E., 'Application of the translinear principle in digital circuits', *IEEE Journal of Solid State Circuits*, **SC-13**, 528–530 (1978).
[11] Seevinck, E., 'An uncommitted integrated circuit', *Trans. S. Afr. Inst. Electr. Eng.* **67** (8), 222–225 (1981).
[12] Morant, M. J., 'Low cost silicon device fabrication in degree courses', *this volume*, p. 2.
[13] Poole, K. F., 'A novel university approach to teaching microelectronics', *this volume*, p. 11.
[14] Current, K. W., 'Undergraduate instruction in bipolar integrated circuit design and fabrication using "Masterslice" integrated circuits', *IEEE Transactions on Education*, **EC-23**, (2) (1980).
[15] Middleton, I. F., 'The design of circuit building blocks using the uncommitted integrated circuit', *MSc(Eng) dissertation, University of the Witwatersrand, Johannesburg* (1977).
[16] Comerford, R. W., 'Users want easier tests for VLSI', *Electronics*, 89–90 (October 20, 1981).
[17] Smith, K. C., *IEEE Transactions on Education*, **EC-22**, (2) 47 (1979).

8

LIQUID CRYSTAL DISPLAYS AS A TOPIC FOR UNDERGRADUATE LABORATORY WORK

B. LAWRENSON
Department of Electrical Engineering and Electronics, University of Dundee, Scotland

INTRODUCTION

When teaching physical electronics topics to undergraduates one is well aware of the fact that, before the operation and design of most electronic and optical devices can be fully appreciated, a relatively large body of fundamental knowledge must have been absorbed. Care is required if such a course is not to become tedious, and the requirements on the lecturer increase if the class consists of students with widely-differing interests. We have found that students can appreciate the importance of understanding how devices work much more readily if they have encountered laboratory work in which they manufacture and test them. It seems that even the most committed Systems man will enjoy giving birth to a transistor, and, having done so, will wonder if, and by what means, its performance might have been improved.

One device which has proved to be popular in this context is the twisted nematic liquid crystal display (LCD), which is fairly easy to make and has a variety of properties which may be investigated with regard to its use in domestic and scientific equipment. Most of these investigations will be suggested by the student himself, once he has been shown how the LCD is made, and so the device lends itself to open-ended or project-type laboratory work. This demonstrates the role that research and development play in the evolution of a commercial product from a physical phenomenon and gives a healthy respect for the quality of LCDs currently available.

Initially the LCD was suggested to fourth-year honours students as a topic for project work, extending over two terms and drawing heavily on papers and conference proceedings for inspiration. Latterly, an experiment which lasts for three afternoons has been included in the third-year laboratory course at Dundee.

The mode of operation, construction, and investigation of a simple LCD test cell are described below, together with details for producing a working seven-segment numerical display.

LIQUID CRYSTALS

Liquids are usually depicted as having no long-range order in either the positioning or the orientation of their molecules. In a relatively small number of cases, however, the orientation of the molecules exhibits a high degree of regularity. (For instance, the molecules may be long dipoles whose axes are all lying in the same direction to within $\pm 20°$.) Liquids which exhibit such a structural feature are termed liquid crystals.

Liquid crystals may result from the mixing of different liquids, when they are termed lyotropic, or they may occur as the initial phase when an organic solid melts. In this case they are called thermotropic. This latter class comprises three types — smectic, nematic and cholestric.

Nematic liquid crystals have their molecules arranged like matches in a box, and in cholesteric material, the direction of molecular alignment changes progressively from one plane to the next, giving a helical structure. It is these two types that have most relevance to current displays.

Because the molecules are polar, their orientation can be altered using an applied electric field. Because the molecules have an elongated shape, the degree to which they scatter incident light depends upon the angle of incidence between the light wave and the molecular axis. An applied electric field is thus able to influence the appearance of the liquid crystal material.

PRINCIPLE OF OPERATION OF A TWISTED NEMATIC LIQUID CRYSTAL DISPLAY

The principle of this type of LCD may be understood with reference to the test cell shown in Fig. 1. The liquid crystal layer shown in Fig. 1(b) consists of nematic material, but it has a progressive twist imposed upon its structure as a result of the way the cell is made. The extent of this is to produce an overall change of 90° in the alignment direction. Light incident upon the polaroid

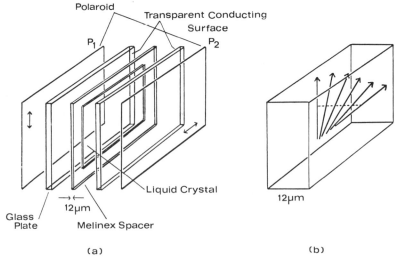

FIG. 1 (a) *Exploded view of a liquid crystal display test cell.* (b) *Structure of the twisted nematic liquid crystal layer.*

sheet, P_1, in Fig. 1(a), is plane-polarized and falls upon the liquid crystal layer. In traversing this layer, the plane of polarization is rotated through 90° by the guiding action of the molecular structure, and the light is freely transmitted by the second polaroid sheet, P_2. The cell is therefore transparent to light, except for the inherent absorption of the two polaroid sheets.

If, now, an electric field is applied to the cell by connecting a p.d. of, say, 10V to the transparent conducting surfaces, the molecules in the liquid crystal layer tend to align with the field and the guiding effect is disrupted. In this case the plane of polarization is not rotated and the light traversing the cell is extinguished by P_2. As shown later, it is by arranging that the electric field is applied to some areas but not to others, that characters are made to appear on the display.

CONSTRUCTION OF AN LCD TEST CELL

Most of the experimental work undertaken by students investigating the twisted nematic type of display can be done using the test cell described in this section. A suggested list of suppliers is given in the Appendix.

The principal materials required are:
(a) Two pieces of linearly-polarizing polaroid sheet.
(b) Two pieces of glass, each with a transparent conducting coating on one face, about $25 \times 20 \times 2$ mm.
(c) Small sheets of Melinex (Mylar) in thicknesses of about 6 μm, 12 μm, 24 μm, 36 μm.
(d) Solution of 0.2% by weight of polyvinyl alcohol (PVA) in de-ionized water.
(e) Liquid crystal material.

None of the materials involved presents a safety problem provided sensible laboratory practice is adhered to and there is reasonable ventilation. Some liquid crystals might conceivably be metabolised to give compounds resembling benzidine, a known carcinogen, and so hands should be washed carefully after the test cell has been made, or lightweight disposable gloves might be provided. However, no carcinogenic or mutagenic activity has been reported for either liquid crystals or photoresist (used in making the working display) by the manufacturer concerned.

To make the test cell the following procedure is adopted:
(i) The pieces of glass are thoroughly cleaned by washing in isopropyl alcohol followed by a 50% solution of methyl alcohol and de-ionized water.
(ii) The conducting faces (electrodes) are identified by resistance probe and marked, and then the pieces of glass are dipped into the PVA solution and allowed to drain dry.
(iii) The PVA covering the electrodes is now rubbed with a paper tissue as shown in Fig. 2. This gives a keyed surface rather like a ploughed field, which dictates the alignment of the molecules in the liquid crystal layer at the electrode interface. The direction of rubbing is marked.

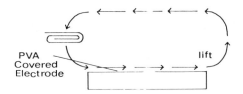

FIG. 2   *Preparation of electrode surface.*

(iv) Two strips of the 12 μm thick Melinex are cut to act as spacers.
(v) The cell is assembled as shown in Fig. 3 and is held together either using bulldog clips or using a purpose-made jig.
(vi) One or two drops of the liquid crystal to be investigated are placed at the open edge of the cell. These are drawn into the cell by capillary action, and the keying action of the PVA coating inside the cell ensures the required 90° overall twist.
(vii) Crocodile clips or edge connectors are used to provide electrical connection to the electrodes.

EXPERIMENTAL WORK

A range of properties may be investigated using the cell described above. Those which we have examined concern the following — all of which are relevant to the commercial success of a working display:
(a) The thickness of the liquid crystal layer
(b) The frequency of the applied field
(c) The angle at which the cell is observed
(d) The temperature
(e) Response times
(f) Different liquid crystals

(a) *Effect of changing cell thickness*

The relationship between the thickness of the liquid crystal layer and the saturation voltage, which is the potential difference required to reduce the transmission of the cell to 20% of its maximum value, may be investigated. We have done this using a photomultiplier and a source of chopped white light, but for economy a silicon photodiode and an oscilloscope could be used. Typical

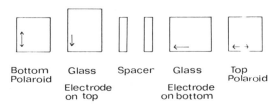

FIG. 3   *Order of assembly of test cell.*

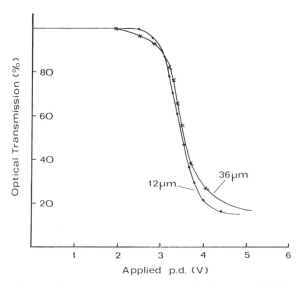

FIG. 4 Plot of optical transmission against applied potential difference showing the insensitivity of threshold voltage to cell thickness for K15 liquid crystal.

results are shown in Fig. 4 for K15 material. It is clear from these results that the saturation voltage is about 4.0 V and is virtually independent of thickness over the range depicted. It would appear that increasing the thickness has the effect of decreasing the restoring forces within the liquid crystal and that this effect compensates for the accompanying reduction in the applied field. An analogous situation is found in the torque required to twist the mid-point of a piece of catapult elastic, which also decreases as the length increases. This suggests that the molecules most affected by the applied field are those in the centre of the layer.

The thickness of the liquid crystal layer is also important in relation to the wavelength of the light. This is because many liquid crystals are birefringent and the light falling on P_2 will then be elliptically polarized, with the degree of ellipticity depending upon the thickness, d, the wavelength, λ, and the two refractive indices, n_E and n_O.

The effect can be demonstrated by making a cell with a wedge-shaped liquid crystal layer, extending in thickness from zero to, say, 12 μm. When this is illuminated with white light a series of coloured bands are seen which demonstrate the expression for transmission, T, given by Gooch and Tarry[1]:

$$T = \frac{\sin^2[\pi/2(1+x^2)^{\frac{1}{2}}]}{1+x^2} \quad \text{with} \quad x = \frac{2d(n_E - n_O)}{\lambda}$$

(b) *Frequency of the applied field*

If an alternating potential difference of variable frequency is applied to the electrodes of a test cell and the transmission is measured, then a limitation on

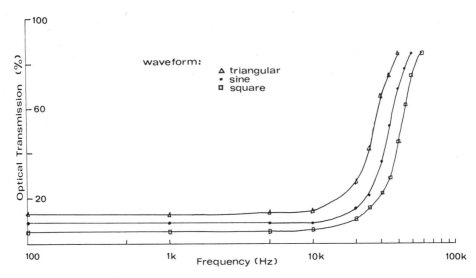

FIG. 5 *Response of the test cell to an alternating potential difference of varying frequency.*

driving frequency becomes apparent, as illustrated in Fig. 5. Above about 20 kHz the inertia of the molecules is too great to permit any significant displacement in the applied field and the cell ceases to switch to the opaque mode at the threshold voltage. This is important in a commercial display, in which an alternating driving voltage is used to eliminate some of the causes of early failure.

(c) *Angular dependence of optical transmission*

Observation of the activated test cell from various angles of elevation, β, and azimuth α, shows that there is a variation of optical transmission with orientation. The effect is discussed by Van Doorn and Heldens[2] and may be understood by considering the orientation of molecules lying in the liquid crystal layer and which have been deflected by the applied field. In the two extreme cases there will be one direction of observation which is parallel to the molecular axes and another which is normal to them. For these cases, the recorded transmission will be found to be quite different. (This angular dependence can be seen if the LCD of a watch or calculator is rotated below a reading lamp.) Measurements have been made of this effect using a modified prism spectrometer table. The test cell was placed upon a mirror on the table and illuminated normally with parallel chopped white light. The spectrometer telescope was replaced by a mounting for the photomultiplier which allowed variation in β. Variation of α was accomplished with the normal rotation of the instrument. Results for a test cell containing a 12 μm layer of E7 are shown in Fig. 6. With 4V applied to the cell, the intensity measured was consistently

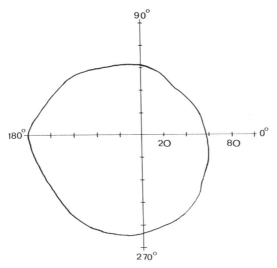

FIG. 6 Polar diagram showing normalized transmission against azimuth, α, for a constant elevation, β, of 60°.

largest for α equal to 225° and least for α equal to 45°. The ratio of these was a maximum for β close to 60°. The effect is voltage-dependent.

Confusion can occur if laser light is used, since the intensity from a polarized laser will vary if the polaroid in the cell is being rotated. Unpolarized lasers are really randomly polarized, and the plane of polarization may change every 5 seconds or so, producing erratic fluctuation in the output.

(d) *Temperature*
Being thermotropic, the liquid crystals used for twisted nematic displays are sensitive to temperature and the most satisfactory materials have a useful temperature range of about $-10°C$ to $60°C$. Other than to observe the changes in appearance which occur at the upper and lower limits, we have not investigated temperature effects. An account of student experiments involving temperature effects in liquid crystals has been given by Fergason[3] who explains the use of cholesteric material to perform thermal mapping, and by Jeppesen and Hughes[4] who discuss a method for detecting the change of birefringence with temperature.

(e) *Response times*
The response time of LCDs is relatively poor, which limits their application, and depends upon the type of material, the temperature, and the thickness. If the test cell is activated by voltage pulses with a pulse height above the threshold value and a duration of around 2s, the transmission of the cell can be recorded on a chart recorder. It is apparent that the relaxation time is about five times the rise time and this is because the internal forces which restore the

structure of the liquid crystal are much smaller than the applied field. If too large a field is applied, the resulting molecular displacements are big enough to produce an oscillation when the field is removed. We have found that a marked improvement in relaxation time can be produced by applying a reverse voltage pulse of, say, 70 μs duration and 5V magnitude at the instant the activating pulse turns off. The unassisted relaxation time of about 0.55 s for a 12 μm thick layer of K15 liquid crystal has been reduced by this technique to about 0.1 s.

(f) *Use of different liquid crystals*

A list of currently-available materials, together with their properties, may be obtained from the manufacturers given in the appendix. Both pure liquid crystal material and specially-formulated mixtures may be obtained. Cells made with pure material suffer from areas of reverse-twist and so it is usual to incorporate a small proportion of a cholesteric liquid crystal to counteract this tendency.

Our investigations have all been done with two mixtures. One is a mixture of the liquid crystal K15 with 0.1% of the cholestric C15 and the other is the proprietory mixture E7, also with 0.1% C15.

MAKING A WORKING NUMERIC DISPLAY

The construction of a working seven-segment display forms a very satisfying conclusion to the range of experiments performed using the test cell. Alternatively it can be offered as a project in its own right.

To make the display, the electrodes of two fresh pieces of conducting glass must be etched to produce patterns such as those shown in Fig. 7. This is accomplished by the standard process of coating with photoresist and contact printing onto the resist through a photographic negative of the required

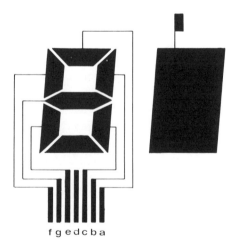

FIG. 7 Seven-segment and back-plane patterns for the electrodes of a working display.

pattern, using ultra-violet light. Developing the resist results in a protective covering over the areas of electrode that are required. The remaining areas are etched away using 0.1 N hydrochloric acid, but the electrode must be wetted and sprinkled liberally with zinc powder before immersion in the acid. During etching, the reduction of the indium and tin oxides in the conducting layer to the metal state, which is then removed by the acid, can be observed.

The remaining resist is then removed, and the display is assembled in the same manner as that used for the test cell.

A photograph of a completed display is shown in Fig. 8. The display may either be manually operated, or activated using one of the many available LCD driver chips. A suitable circuit for counting the output from a low frequency generator is shown in Fig. 9. Input pulses from a function generator are counted by the T.T.L. counter 7490, whose b.c.d. output is decoded into seven-segment format by the CMOS IC 4543B. The CMOS multivibrator 4047 provides an 80 Hz drive signal.

CONCLUSION

Investigating the properties of twisted nematic liquid crystal displays and making working displays have proved to be both popular and worthwhile in

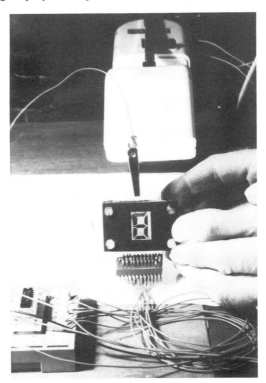

FIG. 8   A working seven-segment display with the segments activated via a modified edge-connector.

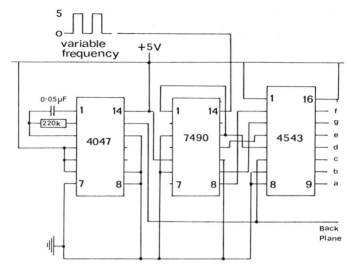

FIG. 9 A suitable driving circuit for the working display.

both formal laboratory and in project work at third and fourth year undergraduate level. Useful results may be obtained with simple equipment, and there is ample scope for students to make their own suggestions. At this level, in contrast to work on the physics of solid state devices, there does not appear to be much scope for the application of theoretical relationships.

ACKNOWLEDGEMENTS
A lot of valuable advice and encouragement has been given by Dr E. P. Raynes and Mrs J. Constant of R.S.R.E., Malvern. The results presented were obtained by Mr M. C. F. Davidson, now with the RAF, and Mr R. McKay, now with Ferranti Ltd., Edinburgh, who also suggested the circuit of Fig. 9.

APPENDIX: SUPPLIERS OF MATERIALS
Liquid Crystals, P.V.A. and solvents: BDH Chemicals Ltd., Poole, Dorset BH12 4NN
Conducting glass: Saunders Roe Developments Ltd., Millington Road, Hayes, Middlesex UBB 4NB
Linear polaroid: A. Gallenkamp & Co. Ltd., P.O. Box 290, Technico House, Christopher Street, London EC2P 2ER
Melinex: I.C.I. Ltd., Plastics Division, Welwyn Garden City, Herts.

REFERENCES
[1] Gooch, C. H. and Tarry, H. A. *J. Phys. D.*, **8**, 1575–1584 (1975).
[2] Van Doorn, C. Z. and Halderns, J. L. A. M. *Phys. Lett.* **47A(2)**, 135–136, (1974).
[3] Fergason, J. L. *Am. J. Phys.*, 38(4), 425–428 (1970).
[4] Jeppesen, M. A. and Hughes, W. T. *Am. J. Phys.*, **38(2)**, 199–201 (1970).

Further reading
Von Willison, K.: *Non-Emissive Electro-Optic Displays* (Plenum, 1975).
Liquid Crystals: free from BDH Chemicals Ltd.

Part 2

AVAILABLE COMPONENTS

9

RECENT DEVELOPMENTS IN THE DESIGN OF MICROCOMPUTER SYSTEM COMPONENTS

P. G. DEPLEDGE
Microprocessor Engineering Unit, Department of Electrical Engineering and Electronics, University of Manchester Institute of Science and Technology, England

1 INTRODUCTION

A previous paper[1] provided a review of the microprocessors available in 1979. During the past two years significant advances in semiconductor production technology have resulted in many sophisticated new microprocessor system components. It is interesting to note that relatively few completely new microprocessors have been announced or produced during this period. For example, almost all the currently-available 16-bit microprocessors were mentioned in Reference [1]. However, developments in the families of components either based upon or to support the 16-bit microprocessors have continued at a rapid pace. Many of these developments are aimed at increasing the computing power of the microcomputer systems by the provision of memory management units, virtual memory support, co-processors and 'intelligent' peripheral controllers. Another significant trend in microprocessor systems, generally, has been the ever-increasing use of CMOS technology to reduce power consumption whilst, through improvements in CMOS device design, equalling the speed of more traditional NMOS components.

The density of memory components continues to increase according to Moore's law with the availability of 64k-bit dynamic read/write memories (RWMs) and 256k-bit read only memories (ROMs). Again memory component designers are making increasing use of CMOS in their designs and, for example, 2k by 8-bit static RWM devices are now available with CMOS memory cells and NMOS interface circuitry. These devices are relatively fast (e.g. 150 ns access time) but have a very low standby power consumption. Perhaps the most significant development in memory design has been the production of electrically erasable and programmable ROMs (EEPROMs) that match industry standard UV-EPROMs in density. Current EEPROMs require 28 volt supplies whilst being programmed. However, some companies have announced that 5 volt-only versions will be available shortly. The inclusion of an EEPROM in a single-chip microcomputer is an exciting prospect and will undoubtedly lead to many novel applications.

There have also been some interesting developments in software aids for microprocessor applications. Many of these are a direct result of the improve-

TABLE 1 *Recent 8-bit single chip microcomputers.*

Manufacturer	Type Number	Technology	Max. memory size		Max. No. of I/O Lines	Serial I/O?	Expandable?
			ROM	RWM			
Intel	8051	NMOS	4k	128	32	Yes	Yes
Motorola	6801	NMOS	2k	128	31	Yes	Yes
Motorola	6805	NMOS CMOS	3·8k	112	24	No	Yes
National Semiconductors	8073	NMOS	Not Applicable	0	0	Yes	Yes
NEC	µCOM-87	NMOS CMOS	6k	128	48	Yes	Yes
Texas Instruments	7000	NMOS	4k	128	28	No	Yes
Zilog	Z8	NMOS	4k	128	32	Yes	Yes

ments in microprocessor architecture to support modern block-structured high-level languages, to provide operating system primitives and to allow re-entrant and position-independent code. The processor which best illustrates the influence that recent developments in software engineering have had on microprocessor design is the Intel iAPX-432 chip-set. It is claimed that this

TABLE 1 *Continued.*

Language Support	Comments
Macro-Assembler. PL/M-51. Tiny-Basic	8k ROM version proposed. CMOS version proposed. Multiply & divide instructions included.
Macro-Assembler.	Enhanced Motorola 6800 instruction set. Multiply instruction included. CMOS version proposed.
Macro-Assembler.	Simplified Motorola 6800 instruction set.
Tiny-Basic	Not a true single-chip microcomputer since external ROM/RWM needed. Directly interprets Tiny-Basic.
Assembler.	Many Z-80-like features. One version has on-board ADC. 64-pin QUIP.
Assembler. Micro-code Assembler.	Microprogrammed. Multiply instruction included.
Assembler.	On-board RWM used as 8 sets of 16 registers. Tiny-Basic version available.

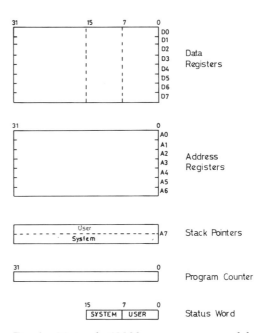

Fig. 1 Motorola 68000 programmers model.

processor directly supports many of the features of the ADA programming language whose development has been supported by the U.S. Department of Defense (DoD). It is intended that ADA will become the standard language for DoD projects and it is hoped, by the DoD, that its use will spread to many civilian projects.

Some of the developments in microcomputer component design will now be considered in more detail. A recent survey which gives more detail on available microprocessors and their support devices is provided in References [2] and [3].

2 MICROPROCESSORS

In Reference [1] the computational power of available microprocessors was illustrated in the form of a spectrum moving from 4-bit single-chip microcomputers, through 8-bit single-chip microprocessors to 16-bit devices. This representation is still valid and we shall now move through this range of devices mentioning many of the major developments in each category.

The 4-bit single-chip microcomputers are by far the most commonly used microprocessor-based devices since they are embedded in an increasing range of consumer goods. For example, over 40 million Texas Instruments TMS1000 series devices were sold in 1981 alone. Originally PMOS devices, the TMS1000 family is now available in both NMOS and CMOS and, in their cheapest form, sell for less than $1 in quantities of over 100,000. Observing the sales that could be achieved with single-chip microcomputers, many manufacturers developed 8-bit devices. These have now reached their second generation and Table 1 lists the features of some of the more recent examples. Note that the use of high-level languages is becoming increasingly important even at this level with the availability of a PL/M compiler for the Intel 8051 and BASIC interpreters for the National Semiconductors 8073 and the Zilog Z8. The Texas TMS7000 is the most revolutionary 8-bit single-chip microcomputer, in that it is microprogrammed and, furthermore, large-volume users are able to tailor the instruction set to their own requirements.

Developments in the 8-bit single-chip microprocessor domain have been fairly limited. The Motorola 6809, now that it is finally in volume production, appears to set the standard for other 8-bit devices to meet. This processor retains many features of the Motorola 6800 and, at the source program level, is compatible with its predecessor. However it has additional pointer registers which provide the addressing modes shown in Table 2. This powerful set of addressing modes make the Motorola 6809 particularly well-suited to running high-level languages such as PASCAL. Another interesting feature of the 6809 is that the program counter relative addressing mode permits the generation of position-independent code. This enables Motorola to sell mass-produced firmware in ROMs that can reside anywhere in the 64k address space. In some simple benchmark tests[2] the 6809 was 2.7 times as fast as the equivalent speed 6800, required 42% fewer instructions and used 33% less code. A CMOS version of this processor is promised by the Japanese company Hitachi. The Zilog Z-80 remains an extremely popular 8-bit microprocessor with sales of approximately 8 million devices in 1981. To maintain this momentum several enhanced versions are beginning to appear. Zilog themselves have announced that an object code compatible processor with multiply and divide instructions, further 16-bit instructions, a 4M byte address space and higher speeds will be

TABLE 2 Motorola 6809 addressing modes. EA = effective address; R = pointer register; A, B, D = accumulators; n = 1 for byte, 2 for word; ( ) = contents of.

Mode	Variations	Operation
Register Direct	Accumulator direct	EA = A,B,D
	Pointer register direct	EA = R
Pointer Register Indirect	Register indirect	EA = (R)
	Post-increment	EA = (R), R ← R + n
	Pre-decrement	R ← R − n, EA = (R)
	Register indirect with offset	EA = (R) + offset
	Register indirect with index	EA = (R) + A,B,D
	Program counter relative	EA = (PC) + offset
Direct	Direct Page	EA = (DP) + offset
	Long	EA = 16-bit address
Indirect	Indirect	EA = ((R))
	Post-increment	EA = ((R)), R ← R + 2
	Pre-decrement	R ← R − 2, EA = ((R))
	With offset	EA = ((R) + offset)
	With index	EA = ((R) + (A,B,D))
	Program counter relative	EA = ((PC) + offset)
	Memory	EA = ((16-bit address))
Immediate Data	—	Data = next byte(s)
Implied	Implied register	EA = A,B,D etc.

TABLE 3 Some 16-bit microprocessor characteristics.

Manufacturer & part number	Data Sizes	Address Range	Virtual Memory?	Supervisory Mode?	Co-processors	High-level Languages Supported	Major Operating Systems	Second Sources	Comments
Intel iAPX 86/10	8,16	1M byte (segmented)	No	No	Yes	PL/M PASCAL FORTRAN 77 BASIC JOVIAL	RMX 86, CP/M-86	Siemens Harris AMD NEC Mitsubishi Fujitsu	CMOS version promised from Harris. Higher performance iAPX 186, 286 due out mid-1982
Motorola 68000	8,16,32	16M byte	No[1]	Yes	No[2]	PASCAL FORTRAN C	VERSADOS, UNIX	Hitachi Rockwell Mostek Signetics Thompson-CSF	1. 68010 will support virtual memory. 2. 68020 will support co-processors.
National Semiconductors NS 16032	8,16,32	16M byte	Yes	Yes	Yes	PASCAL	?	Fairchild Synertek	Samples available late 1981.
Zilog Z-8001	8,16,32,64	8M byte (segmented)	No[3]	Yes	No[4]	PASCAL C PLZ	ZEUS (UNIX)	AMD SGS	3,4 Later versions i.e. Z8003, Z8004 will support these. Non-segmented version (Z8002) available with 64k address range.

available in late 1982. National Semiconductors have also produced a CMOS version of the Z-80 which is now established in volume production, the NSC 800. An Intel 8085 type of multiplexed address/data bus is used on the NSC 800 so that whilst the device is fully compatible with the Z-80 instruction set, it is not pin-compatible.

To bridge the gap between the 8- and 16-bit machines some manufacturers have developed versions of their 16-bit machines with 8-bit external data busses. These devices almost match the computational power of their 16-bit relatives whilst using cheaper 8-bit bus hardware. Intel have employed this philosophy in producing the iAPX 88/10 (or 8088 as it was formerly known). This processor is 100% software compatible with the 16-bit iAPX 86/10 but uses a slightly different multiplexing technique to generate an 8-bit external data bus. The iAPX 88/10 has recently been chosen by IBM for use in its personal computer and thus seems certain to be a popular device with a good selection of applications and systems software. NEC and Fujitsu of Japan have already decided to second-source the iAPX 88/10 and a CMOS version is also promised from Harris Semiconductors.

Five major U.S.-based microprocessor manufacturers, Intel, Motorola, National Semiconductors, Texas Instruments and Zilog, have invested heavily in the development of 16-bit microprocessors and the resulting machines have attracted much attention recently. Table 3 summarises the main features of several of the more recent 16-bit processors and it is worth considering two of these in more detail, the Motorola 68000 and the National Semiconductors NSC 16032.

A programmer's model of the MC68000 is shown in Fig. 1. From this it can be seen that the processor has a 32-bit rather than 16-bit internal architecture. The machine has 8 data registers (DO to D7) which may be used to perform 8, 16 or 32 bit operations and 7 general-purpose address registers. The address registers can be used as base registers, stack pointers and, as with the data registers, may be used as index registers. The 68000 also has two stack pointers, one for the user state and the other for the supervisor state. Whilst in the supervisor state certain privileged instructions can be executed which are not available in the normal user state. Although capable of generating 32-bit addresses, the 68000 only provides a 24-bit external address bus permitting access to a non-segmented 16M byte address space. The addressing modes provided are listed in Table 4. These modes, combined with 56 basic instruction types gives the processor the power of a mid-range minicomputer with excellent support for high-level languages. Two instructions which are particularly relevant to block-structured high-level languages are link (LINK) and unlink (UNLK) which are used to provide data areas on the system stack for nested subroutines, linked lists etc. Fig. 2 shows how these two instructions are used with a subroutine call. From the hardware viewpoint, the 68000 is packaged in a 64-pin dual-in-line package and, unlike most other devices, uses separate address and data busses rather than multiplexing. It allows asynchronous data transfers and so may be interfaced to memory devices of any speed and also

TABLE 4 *Motorola 68000 addressing modes. EA = effective address; AN = address register; DN = data register; RN = address or data register; n = 1 for byte, 2 for words and 4 for long words; ( ) = contents of.*

Mode	Variations	Operation
Register Direct	Data register direct	EA = DN
	Address register direct	EA = AN
Address Register Indirect	Register Indirect	EA = (AN)
	Postincrement	EA = (AN), AN ← AN + n
	Predecrement	AN ← AN - n, EA = (AN)
	Register Indirect with offset	EA = (AN) + offset
	Register Indirect with index and offset	EA = (AN) + (RN) + offset
Direct	Short	EA = (16-bit address)
	Long	EA = (32-bit address)
Program Counter Relative	Relative with offset	EA = (PC) + offset
	Relative with index and offset	EA = (PC) + (RN) + offset
Immediate Data	Immediate	Data = Next Word(s)
	Inerent	Data in instruction
Implied	Implied register	EA = SR, USP, SP, PC

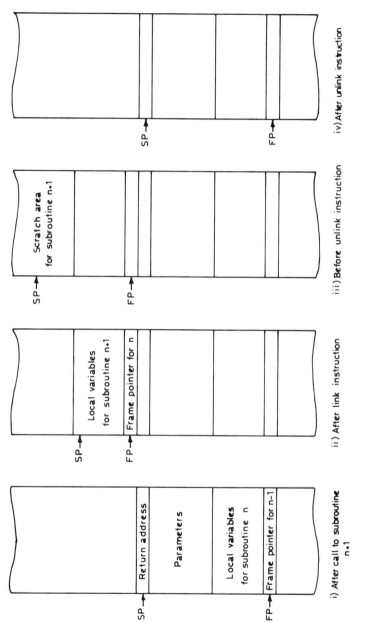

Fig. 2   Link and unlink instructions for the Motorola 68000.

supports synchronous transfers for compatibility with existing Motorola 6800 family components.

The Motorola 68000 represents a very ambitious design for a single chip microprocessor and seems destined to become a popular choice for 16-bit microcomputers. However, the National Semiconductors NS 16032 appears to offer an even more powerful basis for a microcomputer and incorporates many of the best features of the other 16-bit processors. The programmer's model of the NS 16032 is shown in Fig. 3. As can be seen, the processor has 8 32-bit general purpose registers which can be used as pointers or data registers. In addition there are a further 8 dedicated registers of which some deserve a brief description. There are two stack pointers for the user and system stacks although, like the Motorola 68000, only one is valid at any one time. The frame pointer register, FP, is used to reference data on the stack and simplifies the implementation of block structures languages. The 16-bit wide module register, MOD, is used in modular programs to point to an entry in a module table. This module table consists of 3 32-bit descriptions which point to (i) variables that are external to the current module, (ii) the first entry in a link table for the module, and (iii) the instructions in the module. A fourth 32-bit word is reserved for future expansion. The link table for a module is used to reference data or other programs that may be called by the module. The static base register, SB, is essentially a general purpose pointer register similar to those provided on the Motorola 68000. Finally, the interrupt base register contains a pointer to a table which contains pointers to the various interrupt and trap routines. The table can be placed anywhere in memory since the interrupt base register is 24-bits wide.

The NS 16032 has some very interesting address modes in addition to the usual immediate, register and direct modes. These are illustrated in Fig. 4. Note that the scaled indexed addressing mode is primarily intended for accessing arrays and may be used with any of the other address modes except immediate and scaled indexed addressing. Many of the NS 16032 instructions have

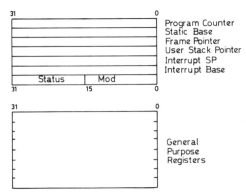

Fig. 3   National Semiconductors NS 16032 programmers model.

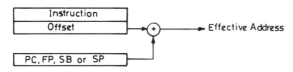
a) Relative register

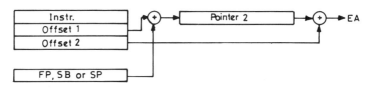
b) Memory relative

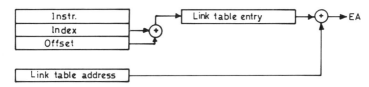
c) External

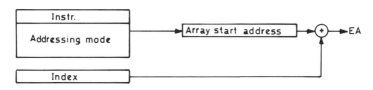
d) Scaled indexed

Fig. 4 National Semiconductors NS 16032 addressing modes.

equivalents on other 16-bit processors but some are unique and warrant a special mention. For example, as well as providing a divide instruction, the 16032 also includes a 'remainder' instruction to enable the remainder to be recovered after a division. The division instruction also permits either truncation towards negative infinity (i.e. -250 divided by 40 yields -7) or towards zero (i.e. -250 divided by 40 yields -6). Most computers only provide the latter. An example of an unusual control order is the CASE instruction. This causes an operand to be added to the program counter and is obviously intended to aid in the implementation of 'CASE OF' or 'SWITCH' statements found in modern high-level languages. The 16032 also includes special instructions to help in the manipulation of strings and multidimensional arrays.

The NS 16032 is packaged in a 48-pin DIL and, unlike the Motorola 68000, uses a multiplexed address and data bus. The processor requires an additional clock generator chip which also provides the read and write control signals. As shown in Table 2, the NS 16032 has been designed from the start with many mainframe-like features, including the support of virtual memory through a complex memory management unit, a floating point unit meeting the proposed IEEE standard for numbers formats and a 5M bytes per second DMA controller. This processor should provide some fierce competition to 'top-end' minicomputers in terms of performance as well as to the other 16-bit processors mentioned here.

Finally, moving away from the essentially single-chip microprocessors, we come to the revolutionary Intel iAPX 432 which is a 3-chip processor, Fig. 5. The iAPX 432 includes many features that have resulted from recent research in computer science. Many of these features are aimed at reducing the development time and maintenance costs of software production by tailoring the hardware architecture to a modern high-level-language environment such as that proposed for the ADA programming language. The prime difference between the iAPX 432 and most other computers is that its operations are defined in terms of manipulating what are called 'objects'. As an example, some simple objects that the iAPX 432 can handle are characters, 16-bit integers, 32-bit real numbers and 64-bit real numbers called primitives. However, because the processor includes system management operations such as inter-process communication, process scheduling and memory management in its hardware, it has to be able to manipulate the data structures associated with these functions. These data structures are called objects. Each object has a set of basic operations defined for it which directly manipulate the data structure and the 432 hardware ensures that these are the only operations that can be

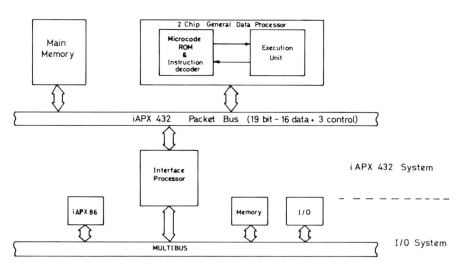

Fig. 5   Intel iAPX 432 system.

Developments in the design of system components 99

TABLE 5 *Typical iAPX 432 objects.*

Object type	Usage	Information content
Processor object	One for each processor in the system	Processor status - e.g. running, halted. Diagnostics and check information. Reference of current process being executed.
Process object	One for each process in the system	Process status - e.g. waiting, running. How the process should be scheduled. Reference to the current context being executed.
Context object	One for each instance of a procedure	Instruction pointer for the context. Stack pointer for the context. Return link to the calling context. References to all objects which may be accessed by the context.
Instruction object	Source of instructions to be fetched and executed.	Contains only instructions - no data.
Data object	Source of data to be manipulated	Integers. Reals. Characters. Combinations of primitives.

employed on the data structure. Thus a significant amount of checking is done automatically by the processor hardware to ensure that invalid operations are not performed. Table 5 lists some typical object types that may be used to make up a program. Note that the 432 is designed for use in multiple processor systems and hence we may need a processor object to describe the state of each physical processor in the system. Even from this brief description of the iAPX 432 it should be clear that the machine is significantly different from existing microprocessors and its many interesting features would require several volumes for a reasonable description. The complexity of the processor is such that Intel suggest that the iAPX 432 is intended for use in projects requiring a software effort of at least 20 man years.

3 MICROPROCESSOR SUPPORT CHIPS

The range of microprocessor support chips has grown dramatically during the past four years. In 1978 a survey of support chips listed approximately 100 devices[3]. This year a similar survey listed over 400 devices. In the same period, the number of different microprocessor types has actually fallen. The motivations for developing so many support devices are due to three factors (i) the so-called 'von Neumann architecture bottleneck', (ii) poor programmer productivity, (iii) the difficulty of interfacing microprocessors to analogue systems. We shall now consider each of these problem areas in more detail.

The von Neumann bottleneck is basically due to the fact that a processor can only execute a single instruction at any one instance. The obvious solution to this problem is to distribute processing tasks amongst separate hardware resources which can operate in parallel. This has led to the concept of the co-processor which is a processor dedicated to one particular set of operations residing on the same bus as the processor and has also led to more sophisticated controller chips. One good example of the former is the Intel 8087 floating point processor. This is designed to operate in conjunction with the Intel iAPX 86/10 and iAPX 88/10 processors and automatically intercepts floating point instructions read from memory by the main processor. It can then execute the floating point instruction whilst the main processor continues execution of other instructions. As shown in Table 3, most 16-bit processors either currently support, or will support, the concept of co-processors. Other support chips which increase system performance by extending the facilities provided by the processor are memory management units (MMU) and direct memory access controllers. The basic purpose of an MMU is to help programmers in co-ordinating the use of the large address spaces available on the new 16-bit microprocessors although some MMUs are used to simply extend the addressing range of a machine. The Hitachi 68451 is an MMU for the 68000 processor family and is said to be more complex than the processor itself. The 68451 allows the description of memory segments as small as 256 bytes and for each segment provides address translation and defines the protection for the segment. Each MMU provides up to 32 segment descriptions. Direct memory address controllers for microprocessors have been available for several years and they have now become very sophisticated For example the Intel iAPX 86/11 (or 8089) although called an input/output processor, is essentially a DMA controller. The iAPX 86/11 permits masked compare operations and data translation during DMA at a rate of 1.25M bytes/sec. The transfer control conditions include byte count, scan-until, scan-while, single transfer and transfer on an external event. Each iAPX 86/11 provides control for two channels and allows the addressing of up to 1M byte of store.

The use of MMUs and DMA controllers eases the burdens of a programmer as well as enhancing performance. Many other more specialised devices are also now available to ease the programming tasks for handling peripheral devices. These now include floppy and Winchester disk controllers, CRT controllers, keyboard and display interfaces, printer controllers and data encryption devices. The specialised functions of these devices are normally invoked by relatively simple commands from the host microprocessor which could be regarded as an extension to the processor instruction set. Many of these peripheral control chips are available from companies other than the mainstream microprocessor manufacturers. For example Western Digital, who are well known for their excellent floppy disk controller chips, also produce data encryption units, data link controllers and serial input/output devices.

Finally we come to analogue input/output devices. There are now a great variety of 8-bit ADCs and DACs with conversion times in the range of 10 to 500 μsec costing between £3 and £10. These devices are compatible with the microprocessor data busses and thus do not require any additional support chips. As an example, Ferranti manufacture a successive approximation ADC (ZN 427) with a 15 μsec conversion time, tri-state outputs and on-chip voltage reference and an R-2R DAC with an 800 nsec setting time, input latches and an on-chip voltage reference. 10 and 12-bit devices are also becoming more popular and will no doubt continue to do so as longer wordlength microprocessors become more predominant.

4 MICROPROCESSOR FIRMWARE

As we have already observed, many of the more recent microprocessors have been designed to support high-level languages and to ease program development generally. It has been possible to buy monitors and interpreters (e.g. for BASIC and FORT) for several years, but these usually have the disadvantage that they are in absolute code form and must reside at predetermined and fixed locations with supporting device drivers also at fixed locations. This position-dependence is a severe constraint and limits the usefulness of much firmware. Recently, with the availability of processors that have extensive program counter relative addressing (e.g. Motorola 6809, Motorola 68000, National Semiconductors NSC 16000), manufacturers are able to produce ROM based programs that are position-independent. Such firmware is only just appearing on the market but will become a major benefit to programmers of the future and will drive down the costs of software production and maintenance significantly.

One example of position-independent firmware is the Motorola 6839 which is a highly optimised implementation of the IEEE proposed standard for binary floating point arithmetic for the Motorola 6809. Several sophisticated monitor/debug packages and interpreters are also available for the same processor in a position-independent form.

Another significant development in the provision of system firmware is the increasing availability of real-time multitasking operating system kernels. An excellent example is the Intel 80130 which is a 16k byte ROM combined with two 16-bit timers, a baud-rate generator and interrupt logic. The 80130 implements over 30 of the most important primitives of the RMX 86 operating system and is available in versions to support the iAPX 86/10 and iAPX 88/10 processors. Such components will become increasingly popular for the new 16-bit processors, most of which will be employed in applications which require a multi-tasking operating system.

5 CONCLUSIONS

The microprocessor is now ten years old and, as this paper shows, has developed from being a crude and simple device with limited processing power to a powerful system component equalling the performance of mini- and even mainframe computers. Yet we have still not reached the limits of NMOS technology, and component integration levels and device speeds will probably continue to increase for several more years. Already Hewlett Packard have produced a single-chip 32-bit processor using 450,000 transistors operating with an 18 MHz clock[4]. The instruction set of this processor provides many of the functions of an advanced mainframe CPU, including floating point arithmetic. The processor can execute a 32-bit integer addition in 55 ns, a 32-bit integer multiply in 1.8 µs and a floating point multiply in 10.4 µs. As well as producing processors which will form the heart of powerful computer systems, it is also clear that manufacturers will continue to develop more advanced single-chip microcomputers. These will form the basis of many innovative and new products and, as electrical engineers, it is likely that we will incorporate these devices in our designs.

The microprocessor now forms the basis of the majority of new electronic designs. It will be interesting to observe if this position remains the same in the face of increasing competition from custom and semi-custom VLSI components during the next decade.

REFERENCES

[1] Depledge, P. G., 'A Review of Available Microprocessors', *I.J.E.E.E.*, **16**, No. 2, pp. 114–123, (1979).
[2] Cushman, R. H., *Eighth Annual µP/µC Chip Directory*, *EDN*, **26**, No. 22, pp. 100–220, (November 1981).
[3] Cushman, R. H., *Fifth Annual µC Support Chip Directory*, *EDN*, **27**, No. 1, pp. 155–206, (January 1982).
[4] Beyers, J. W., Dohse, L. J., Fucetola R. L. K., Lob, C. G., Taylor, G., Zeller, E. R., 'A 32-bit VLSI CPU Chip', *IEEE Journal of Solid State Circuits*, **SC-16**, 5, pp. 537–545, (October 1981).

Part 3

LABORATORY USE OF MICROCOMPUTERS

10

A SINGLE-BIT MICROCOMPUTER EXPERIMENT

R. M. HODGSON and E. J. HAMILTON
Department of Electrical Engineering, University of Canterbury, Christchurch, New Zealand

INTRODUCTION
For the average student, the first contact with the microprocessor can be both fascinating and totally confusing. The experiment described here was developed to reduce the trauma of this first contact and to act as a conceptual bridge between conventional digital circuits and standard eight-bit microcomputers. The experiment has been designed around the Motorola MC14500B single-bit industrial control unit (ICU). This device was developed as a solid-state substitute for conventional relay controllers, and can be thought of as a primitive microprocessor. In the experiment, the basic hardware and software concepts of microprocessor systems are introduced and demonstrated within a typical five-hour laboratory session.

HARDWARE DESCRIPTION
The Motorola MC14500B is a single chip, one-bit, static CMOS processor optimized for decision-oriented tasks. The processor is housed in a sixteen-pin package and features sixteen four-bit instructions. The instructions perform logical operations on data appearing on a one-bit bidirectional data bus and data in a one-bit accumulating results register within the ICU. All operations are performed at the bit level.

Fig. 1 is a block diagram for the device and the instruction set is listed on Fig. 2.

The experiment is based on the use of a modified version of a demonstration system described in the Motorola device handbook[2]. A block diagram for the system is given in Fig. 3. This was assembled on a printed circuit with a perspex cover onto which were mounted programming switches and sockets for I/O ports. Light-emitting diodes are used to show the status of the ports, the clock, the RAM and the results register. In the demonstration system an 'interlaced' memory structure is used, with alternate 4-bit wide locations used to store instructions and the corresponding operand addresses. The least significant address bit is supplied by the clock. In our implementation, the system can either be clocked manually or using an internal slow (1 Hertz) clock.

SOFTWARE DESCRIPTION
The instruction set used in the experiment is shown in Fig. 2. It is the simplicity of the instruction set and the absence of a monitor program that makes the

A single-bit microcomputer experiment

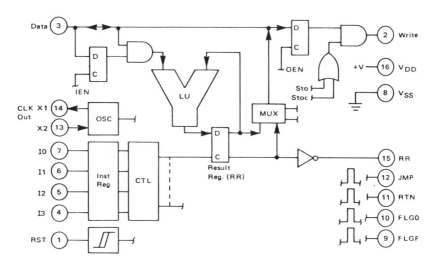

FIG. 1 Block diagram for the Motorola MC14500B Industrial Control Unit.

Instruction Code		Mnemonic	Action
#$_0$	0000	NOPO	No change in registers. R→R, FLG0 Flag Pulsed
#$_1$	0001	LD	Load Result Reg. Data → RR
#$_2$	0010	LDC	Load Complement $\overline{\text{Data}}$ → RR
#$_3$	0011	AND	Logical AND. RR D → RR
#$_4$	0100	ANDC	Logical AND Compl. RR $\overline{\text{D}}$ → RR
#$_5$	0101	OR	Logical OR. RR. + D → RR
#$_6$	0110	ORC	Logical OP Compl. RR + $\overline{\text{D}}$ → RR
#$_7$	0111	XNOR	Exclusive NOR. If RR = D.RR → 1
#$_8$	1000	STO	Store. RR → Data Pin, Write ← 1
#$_9$	1001	STOC	Store Compl. $\overline{\text{RR}}$ → Data Pin, Write ← 1
#$_A$	1010	IEN	Input Enable. D → IEN Reg.
#$_B$	1011	OEN	Output Enable. D → OEN Reg.
#$_C$	1100	JMP	Jump. JMP Flag ← Flag Pulsed
#$_D$	1101	RTN	Return. RTN Flag Pulsed.Skip Next Instruction
#$_E$	1110	SKZ	Skip next instruction if RR = 0
#$_F$	1111	NOPF	No Change in Registers RR→RR, FLGF Flag Pulsed

(a)

#$_0$	0000	RSTO	Stops Program RSTO Flag pulsed
#$_D$	1101	CLTL	Clears O/P 8-15 CLTL Flag Pulsed
#$_F$	1111	RSTF	Loops Program RSTF Flag Pulsed

(b)

FIG. 2 (a) The instruction set (b) Modifications to the instruction set.

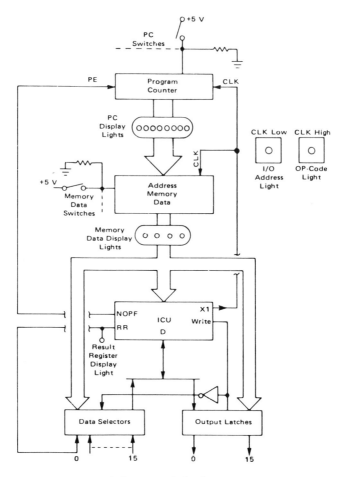

FIG. 3   Block diagram of the demonstration system.

device suitable for our application. A looping control structure is used. That is, with the exception of an unconditional jump to the bottom of RAM, branching and jumping operations are not made available. See the appendix for an elaboration of this point.

PROCEDURE

The experiment was introduced to second-year students. Preparation was not demanded; they worked in small groups and were given intensive tutorial support by a postgraduate demonstrator. The students were provided with the salient information on the MC14500B and the design of the demonstration unit. They were first required to familiarize themselves with the hardware and to run a number of demonstration programs. Finally they were required to design a program to meet the requirements for a controller for a gas fired boiler

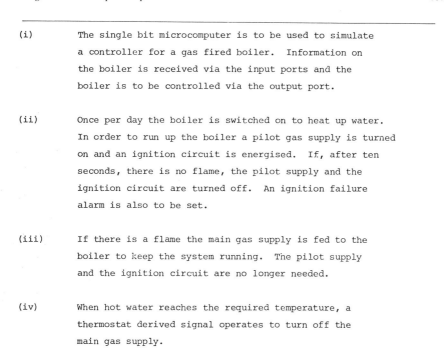

(i) The single bit microcomputer is to be used to simulate a controller for a gas fired boiler. Information on the boiler is received via the input ports and the boiler is to be controlled via the output port.

(ii) Once per day the boiler is switched on to heat up water. In order to run up the boiler a pilot gas supply is turned on and an ignition circuit is energised. If, after ten seconds, there is no flame, the pilot supply and the ignition circuit are turned off. An ignition failure alarm is also to be set.

(iii) If there is a flame the main gas supply is fed to the boiler to keep the system running. The pilot supply and the ignition circuit are no longer needed.

(iv) When hot water reaches the required temperature, a thermostat derived signal operates to turn off the main gas supply.

(v) The demonstration system is to generate control signals to turn the gas supplies on and off, to monitor a thermostat input and to set the ignition failure alarm. An output latch is used as the ignition failure alarm.

FIG. 4   The gas-fired boiler problem.

given in Fig. 4. A flow chart for the problem was readily developed by most students but most required assistance from the demonstrator to arrive at a satisfactory final solution.

DISCUSSION AND CONCLUSIONS

The single-bit microcomputer experiment was introduced to bridge the conceptual gap between simple logic devices and fully-fledged microprocessor systems. As part of a design class held later in the year, the students take a short course on microcomputers. The course is based on the 8085 and includes laboratory sessions in which programs are written and simple interfaces designed and tested (using SDK 85s). In discussion, most students claimed to have derived benefit from their earlier 'hands on' experience with the single-bit microcomputer experiment discussed here.

REFERENCES

[1] Gregory, V., and Dellande, B. *MC14500B Industrial Control Unit Handbook*, Motorola Inc. (1977).
[2] *Ibid.* Ch. 5.

APPENDIX

Modifications to the standard instruction set

To suit the purposes of the experiment, three flag associated instructions have been used to modify the behaviour of the demonstration system.

(i) The RSTF instruction is used to reset the program counter and to reset the latches on output ports 0 to 7 leaving ports 8 to 15 unchanged.
(ii) The RSTO instruction stops the program counter without resetting it. This stops the program on successful execution.
(iii) The CLTL (clear top latches) instruction resets the latches on output ports 8 to 15.

11

ANALOG TO DIGITAL CONVERSION USING A MICROPROCESSOR

A. M. CHADWICK and W. HERDMAN
School of Electronic Engineering, Newcastle upon Tyne Polytechnic, England

1 INTRODUCTION

Today the teaching of microprocessors is finding its way into many degree courses, in particular, those in electronic engineering. It is very important for the student to appreciate the interfacing problems which arise when a microprocessor-based system is employed in practice. These problems are well illustrated by examining systems for analogue-to-digital conversion based around a microprocessor and, at the same time, this offers a new approach to laboratory work in this important area.

One current approach is to use SSI or LSI logic patch boards where the student examines the operation of the converter alone. Unfortunately the student's time in the laboratory consists largely of wiring up the circuitry, and often several laboratory periods are required to allow him to fully appreciate the three basic types: single ramp, tracking, and successive approximation. This article describes a laboratory package which has been successfully used with degree students to demonstrate these three types of A to D converter. It also serves to introduce the student to situations where the microprocessor forms an integral part of the system rather than a stand-alone unit.

The experiments described below were developed in a laboratory equipped with Motorola M6802-based systems, the Creative Micro Systems 9600A single-board microcomputer, but could be carried out with almost any development system.

2 EXPERIMENTAL WORK

2.1 *The open loop test*

Fig. 1 shows the block diagram for the general-purpose converter module used throughout the experimental work. This module can actually be used as either a D/A or A/D converter simply by inserting the link shown.

The student is first asked to write a program which will output two digital words from the microcomputer's parallel port (PIA) to the D/A converter which in turn will force the comparator to switch repetitively so that the output can be observed. Fig. 2 shows the waveforms the student is expected to generate. This is intended to make him aware of the potential problems involved when combining the microprocessor with standard analog circuitry, in this case the poor slew rate of the 741 operational amplifier used for current-voltage

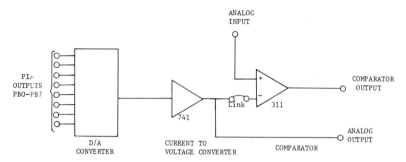

FIG. 1   Module block diagram.

conversion. Attention should be drawn to the fact that this problem may necessitate the inclusion of a small time delay sub-routine in the conversion programs after the application of the conversion byte at the PIA output and before reading the state of the comparator at the PIA input.

Clearly an operational amplifier with a faster slew rate could be used but it is felt that the use of a 741 highlights a potential problem area. The upper trace shows the output of the 741 amplifier while the lower trace shows that the 311 comparator switches very quickly.

The program to demonstrate this aspect, which is essentially an open loop response test, is shown in Table 1.

3   THE SINGLE RAMP CONVERTER

This technique uses an up-counter whose final value is controlled by the comparator output. At the start of conversion the counter is first set to zero and will then increment until the output from the D/A convertor exceeds the analog input on the comparator. This causes the comparator to switch, which

TABLE 1   *Open loop test program.*

E500	7F	CLR		$E3D1		
E503	7F	CLR		$E3D3		
E506	7F	CLR		$E3D0		
E509	86	LDA	A	$FF		PIA
E50B	B7	STA	A	$E3D2		CONFIGURATION
E50E	86	LDA	A	$04		ROUTINE
E510	B7	STA	A	$E3D1		
E513	B7	STA	A	$E3D3		
E516	86	LDA	A	$FF		OUTPUT MAXIMUM COUNT
E518	B7	STA	A	$E3D2		
E51B	8D	BSR		$0B	E528	DELAY
E51D	01	NOP				
E51E	4F	CLR	A			OUTPUT MINIMUM COUNT
E51F	B7	STA	A	$E3D2		
E522	8D	BSR		$04	E528	DELAY
E524	01	NOP				
E525	20	BRA		$EF	E516	REPEAT SEQUENCE
E527	01	NOP				
E528	C6	LDA	B	$02		TIME DELAY
E52A	5A	DEC	B			ROUTINE
E52B	26	BNE		$FD	E52A	
E52D	39	RTS				

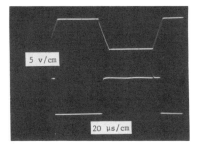

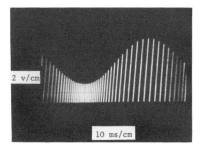

FIG. 2 Open loop test waveforms.   *FIG. 3 Single ramp converter waveforms.*

in turn inhibits the counter. The digital value now existing on the input to the D/A converter represents the analog voltage at the input.

In this experiment the microprocessor replaces the up-counter and the inhibit circuitry. The program uses accumulator A to produce the up-counter via PB0 to PB7 while the least significant bit of accumulator B is used to determine the state of the comparator output via PA0. When the comparator output goes to zero the count is inhibited, the conversion is complete and the program then returns to the start and initiates the next conversion. The program from address E800 to E812 is concerned only with initialising the PIA.

The short delay routine (13 μs) at address E817 is intended to overcome the problem described in section 2.1.

It is an interesting additional exercise for the student to investigate the performance of the system as the delay is reduced. With a d.c. input on the comparator the reduction in time delay can be seen to give an overshoot due to the slow response of the operational amplifier.

The variable conversion time associated with the single ramp A/D converter is very well demonstrated by using a sine wave input (with a d.c. offset to ensure unipolar signals) producing typically the results shown in Fig. 3.

TABLE 2 *Single ramp converter program.*

E800	CE	LDX		$E3D0			
E803	6F	CLR		01,X			
E805	6F	CLR		03,X			
E807	6F	CLR		00,X			PIA
E809	86	LDA	A	£FF			CONFIGURATION
E80B	A7	STA	A	02,X			ROUTINE
E80D	86	LDA	A	$04			
E80F	A7	STA	A	01,X			
E811	A7	STA	A	03,X			
E813	4F	CLR	A			O	SET COUNT TO ZERO
E814	B7	STA	A	$E3D2			OUTPUT TO D/A
E817	8D	BSR		$0A	E823		TIME DELAY
E819	F6	LDA					READ COMPARATOR
E81C	C4	AND	B	$01			MASK
E81E	27	BEQ		$F3	E813		CONVERSION COMPLETE
E820	4C	INC	A			L	INCREMENT COUNT
E821	20	BRA		$F1	E814		
E823	39	RTS				9	

4 THE TRACKING CONVERTER

In a tracking type A/D converter the single up-counter employed previously is replaced by an up/down counter. The condition of the comparator output now dictates the direction of the count so that the D/A converter output can 'track' the analog input. In effect, the starting point for each conversion is the end point of the previous conversion.

The single up-counter program already developed is very easily modified to include a down count and the output from the comparator now dictates the direction of the count. The program of Table 3 shows two extra lines and two altered branch instructions when compared with Table 2. The waveforms generated by this type of converter are reproduced in Fig. 4 which, for the sake of clarity, is only a five-bit conversion, the three least significant bits being discarded.

TABLE 3 *Tracking converter program.*

E825	CE	LDX		$E3D0			
E828	6F	CLR		01,X			
E82A	6F	CLR		03,X			PIA
E82C	6F	CLR		00,X			CONFIGURATION
E82E	86	LDA	A	$FF			ROUTINE
E830	A7	STA	A	02,X			
E832	86	LDA	A	$04			
E834	A7	STA	A	01,X			
E836	A7	STA	A	03,X			
E838	4F	CLR	A			O	SET COUNT TO ZERO
E839	B7	STA	A	$E3D2			OUTPUT TO D/A
E83C	8D	BSR		$0E	E84C		TIME DELAY
E83E	01	NOP					
E83F	F6	LDA		$E3D0			READ COMPARATOR
E842	C4	AND	B	$01			MASK
E844	27	BEQ		$03	E849		
E846	4C	INC	A			L	COUNT UP
E847	20	BRA		$F0	E839		START
E849	4A	DEC	A			J	COUNT DOWN
E84A	20	BRA		$ED	E839		START
E84C	39	RTS				9	

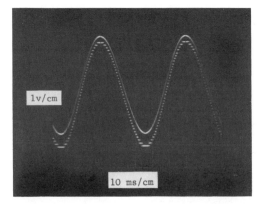

FIG. 4 Tracking converter waveforms.

There is considerable scope for additional work on this converter, in particular, the investigation of the maximum input rates at which tracking can be maintained successfully and the effect of reducing the number of bits in the counter.

It is also interesting to observe the performance of this converter when a square wave input is applied.

5 THE SUCCESSIVE APPROXIMATION CONVERTER

Manifestly, this converter is the most difficult of the three described here for the student to realise if asked to use patchboard or hardwire methods. It is also the most difficult program the student has to develop and cannot easily be derived from the previous programs. However it is suggested that the student be taught the principles of this converter during lectures and asked to develop the necessary algorithm before the laboratory period. Again the conversion byte is contained in accumulator A and transferred to the D/A converter via the PIA. Similarly the comparator output is read into the least significant bit of the B accumulator. An additional test bit is required in this technique which performs two functions, the first being simply to control the number of successive approximations. The second and essential feature of the system is the OR-ing of this test bit with the partly completed conversion byte. The OR process is controlled by the state of the comparator output such that a high comparator output forces an inclusive-OR of the test bit with the conversion byte and hence fixes the bit in the conversion byte; conversely, a low output from the comparator gives rise to an exclusive-OR of the two words and thus removes the test bit from the conversion byte. The flow diagram (Fig. 5) illustrates this algorithm. The successive approximation program and resulting waveforms are shown in Table 4 and Fig. 6. A small delay is included between conversions to highlight the start of conversion.

6 FURTHER WORK

The combination of a D/A converter module and a microprocessor offers a wide range of opportunities for further experimental work. As an example, any repetitive waveform can be easily generated and this approach can be extended to illustrate many techniques in this area. The 'Look-Up' table method of sinewave generation is neatly demonstrated (Fig. 7) by storing the sinewave data in memory and repeatedly outputting this under microprocessor control. This system can be modified to produce a swept frequency sinewave output by the inclusion of a continuously reducing time delay in the output routine.

7 CONCLUSION

The experimental programme described here makes an excellent laboratory package for students of quite widely varying abilities. The counter and tracking converters are both very easy to realise and can be successfully completed in a relatively short time by most students. The successive approximation technique is no doubt more demanding but all three forms of converter can usually be

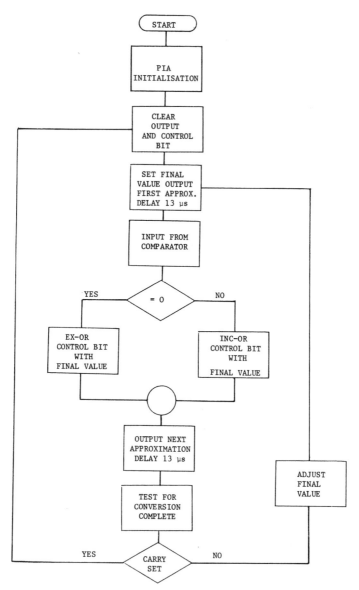

FIG. 5 *Successive approximation converter flow diagram.*

comfortably completed within a normal three-hour laboratory period, if adequate time for software development is allowed.

The microprocessor development system used throughout offers a very good monitor program with the facility to single-step through a program execution as well as displaying the register contents after each instruction. This allows the

Analog to digital conversion using a microprocessor

TABLE 4 *Successive approximation converter program.*

E850	CE	LDX	$E3D0		
E853	6F	CLR	01,X		
E855	6F	CLR	03,X		PIA
E857	6F	CLR 00	00,X		CONFIGURATION
E859	86	LDA A	$FF		ROUTINE
E85B	A7	STA A	02,X		
E85D	86	LDA A	$04		
E85F	A7	STA A	01,X		
E861	A7	STA A	03,X		
E863	4F	CLR A			
E864	B7	STA A	$E3D2		SET OUTPUT TO ZERO
E867	8D	BSR	$26	E88F	DELAY
E869	0C	CLC			CLEAR CONTROL BIT
E86A	86	LDA A	$80		
E86C	97	STA A	$50		SET FINAL VALUE
E86E	B7	STA A	$E3D2		FIRST APPROXIMATION
E871	8D	BSR	$22	E895	DELAY
E873	F6	LDA B	$E3D0		READ COMPARATOR
E876	C4	AND B	$01		MASK
E878	27	BEQ	$10	E88A	
E87A	9A	ORA A	$50		INCLUDE CONTROL BIT
E87C	B7	STA A	$E3D2		OUTPUT
E87F	8D	BSR	$14	E895	DELAY
E881	76	RDR	$0050		ADJUST CONTROL BIT
E884	25	BCS	$DD	E863	NEXT CONVERSION
E886	9B	ADD A	$50		ADJUST FINAL VALUE
E888	20	BRA	$E4	E86E	NEXT APPROXIMATION
E88A	98	EOR A	$50		EXCLUDE CONTROL BIT
E88C	20	BRA	$EE	E87C	
E88E	01	NOP			
E88F	CE	LDX	$0002		TIME DELAY 1
E892	09	DEX			
E893	26	BNE	$FD	E892	
E895	39	RTS			TIME DELAY 2

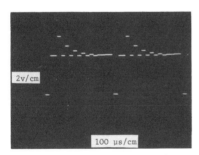

FIG. 6 Successive approximation converter waveforms.

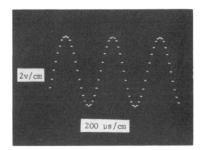

FIG. 7

student to appreciate more fully the operation of each converter, particularly that based on the successive approximation technique.

Finally, if the programs are transferred to an EPROM the whole series of programs can then be used as a demonstration during lectures prior to the students attempting the experiments, by a small program modification each technique can then be used as an EPROM based subroutine in more advanced data acquisition systems.

12

A MICROCOMPUTER-CONTROLLED EXPERIMENT TO MEASURE SEMICONDUCTOR MATERIAL PROPERTIES

K. E. SINGER and H. D. McKELL
Department of Electrical Engineering and Electronics, University of Manchester Institute of Science and Technology, England

1   INTRODUCTION

Most undergraduate electrical engineering and electronics syllabuses now contain a number of courses dealing with the physics of operation of semiconductor devices and the technology of their manufacture. An essential element in these courses is a description of the basic conduction processes in semiconductor materials and how these relate to device characteristics. The primary objective of the laboratory experiment described in this paper was to reinforce lecture material on the more elementary aspects of the subject.

The experiment involves the measurement of the conductivity and the Hall coefficient of a semiconductor as a function of temperature over the range 200 to 400 K. Analysis of the data allows the student to determine the conductivity type, the effective dopant concentration, the magnitude of the energy gap and both the magnitude and the temperature variation of the electron and hole mobilities.

The experiment is designed around a microcomputer which is used both to control the experiment and manage the data collection, as well as perform some initial reduction of the data and produce hard copy graphical output which forms the basis for the further analysis. This approach ensures consistently good data which would be very difficult to achieve by conventional analogue instrumentation in the far-from-ideal conditions of an undergraduate teaching laboratory.

The use of the computer has the additional advantage that it introduces the student to the type of automatic measurement system which is becoming increasingly common in the field of semiconductor material and device assessment — as well as many other areas of measurement. This aspect of the experiment has turned out to be an important secondary objective.

The disadvantages of a computer-based measurement are that it does not allow students an opportunity to develop their own measurement skills, and, more importantly, it can obscure the details of the method.

2 DESCRIPTION OF THE EXPERIMENT

2.1 The Van der Pauw Method

The method used to measure conductivity and the Hall coefficient is that due to Van der Pauw[1]. In this method, four small ohmic contacts are arranged around the periphery of a semiconductor wafer of uniform thickness. In principle, the shape of the wafer and the position of the contacts are arbitrary, but the use of a symmetrical sample considerably simplifies the measurement and the subsequent calculations.

To determine the conductivity, a current is passed through two adjacent contacts (A and B — see Fig. 1) and the resultant voltage across the other two is measured. The simplified form of the Van der Pauw expression for a symmetrical sample yields the conductivity explicitly as

$$\sigma = \frac{\ln 2}{\pi d} \frac{I_{AB}}{V_{CD}}$$

where d is the sample thickness.

The Hall coefficient is obtained by placing the wafer in a magnetic field normal to its plane and passing a current through two opposite contacts. The

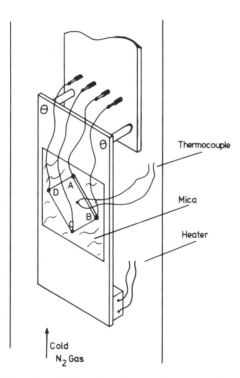

FIG. 1 The sample and its holder. For clarity the spring clip retaining the sample and the thermocouple is omitted.

Hall voltage is measured across the other two contacts and, for a symmetrical sample, the Hall coefficient is found from

$$R_H = \frac{d}{B} \frac{V_{BD}}{I_{AC}}$$

where B is the magnitude of the field.

Even with great care in the preparation of the sample and the attachment of the four contacts it is inevitable that there will be some asymmetry in the geometry. This is particularly true when soldered contacts are used. As long as the asymmetry is small, the simple form of the conductivity expression is adequate — although if high accuracy is required it may be necessary to rotate the current and voltage positions through 90° (i.e. measure I_{BC}/V_{DA}) and average the two values of conductivity.

However, for the Hall measurement, even a small asymmetry will cause a significant misalignment voltage to be present in the absence of a magnetic field. This must be subtracted from the voltage measured in the field giving

$$R_H = \frac{d}{B} \Delta r$$

where

$$\Delta r = \left(\frac{V_{BD}}{I_{AC}}\right)_{\text{in field}} - \left(\frac{V_{BD}}{I_{AC}}\right)_{\text{out of field}}$$

Thus the experiment requires the sample to be taken through the temperature cycle three times; once to measure the conductivity, once to measure the Hall misalignment voltage and finally to measure the Hall voltage. In principle it would be possible, by switching contacts and moving the sample in and out of the field, to collect all the data with a single temperature sweep. However, the added complexity would be considerable and important details of the measurement would be obscured from the student.

A further complication is caused by small temperature differences occurring at the two voltage contacts. This gives rise to an unwanted thermal e.m.f. which is eliminated by measuring in quick succession the voltages with and without a current passing.

2.2 The sample and holder

In order to gain the maximum information from the experiment it is important that the sample exhibits both intrinsic and extrinsic behaviour within the available temperature range. Simple soldered contacts put an upper limit on the temperature of ~ 400 K which precludes the use of silicon. In the experiment n-type germanium with an effective donor density of 5×10^{13} cm^{-3} has been used, but any value in the range 5×10^{13} to 3×10^{14} would be suitable. It is also possible to use p-type material which would give the added interest of a sign change in the Hall coefficient as the sample goes intrinsic.

The sample is in the form of a square plate with 1 cm sides and is approximately 1 mm thick. Adequate ohmic contacts are made by directly soldering fine copper wires to each corner using solder paste (Fryolux). The arrangement of the sample on its holder is shown in Fig. 1. The slice is insulated from the copper base plate and the spring clip by thin sheets of mica. A 25 W soldering iron element, used for heating the sample, is fed from a variac which is adjusted to give a heating rate of approximately 30°/min. Cooling of the sample is achieved by passing nitrogen gas through a copper spiral immersed in liquid nitrogen and then through the glass tube containing the sample. Nitrogen gas is flushed through the tube for a few minutes prior to cooling in order to prevent condensation forming on the sample. The sample temperature is measured by a fine alumel-chromel thermocouple held in contact by the spring clip. The reference junction is immersed in ice.

2.3 Electrical measurements

A schematic of the measurement system is shown in Fig. 2. Current is fed to the sample from a near-constant current source consisting of a 30 V battery in series with a 5 KΩ resistor. The current is switched by a reed relay and the voltage across the 10 Ω series resistor is used to measure its magnitude. The reed relay is energised by the D/A converter which in turn is linked to the computer. This allows the current to be switched off under programme control in order to measure any thermal e.m.f. across the voltage contacts.

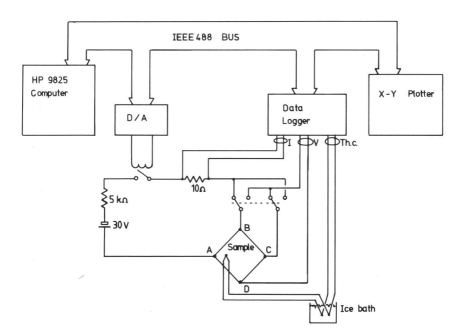

FIG. 2 Schematic of the measurement system.

The sample voltage, the voltage across the current measuring resistor and the thermocouple output are fed to a data logger which itself is linked to the computer. On receiving a trigger signal from the computer, the data logger scans the three channels and outputs the data to the computer.

Interconnection between the computer, the data logger, the D/A converter and the x-y plotter used to display the data, is via the standard IEEE 488 interface bus. The computer used is a Hewlett-Packard type 9825, but virtually any desktop machine with suitable I/O facilities would suffice.

2.4 Data collection and reduction

The procedure for data collection is essentially the same for all three runs. The sample is first cooled and at 200 K the data collection sequence is initiated. With the sample current switched off the voltage is measured and stored. The computer then switches the current on and re-measures the voltage as well as the current and the thermocouple e.m.f. Immediately, the first voltage is subtracted from the second, the conductivity is computed and the result is stored along with the corresponding thermocouple e.m.f. The raw voltage and current data are not retained.

Following this, the computer reads the thermocouple output repetitively, and when the value is greater than 50 µV higher than that from the first data point (corresponding to a minimum temperature increment of $\sim 1.2°$) the measurement sequence is repeated and the second conductivity value and thermocouple e.m.f. stored. This process is repeated until the temperature reaches 400 K when a message is displayed telling the student to switch the heater off.

Following the data collection sequence, the measured thermocouple e.m.f.s are converted to temperature by a spline interpolation routine using ten thermocouple reference table points, which are stored in the programme. The conductivity vs. temperature data is then plotted.

The two sequences for the Hall measurement are the same, but this time the data is plotted as Hall voltage per unit current as a function of temperature. The in-field and out-of-field data are plotted on the same graph to facilitate the required subtraction.

3 THE DATA

Figs. 3(a) and (b) show the data from the experiment as it is produced by the computer. In place of a conventional report, students are required to carry out a number of tasks and answer a series of questions which are designed to lead them through the analysis of the data.

The first two tasks simply require the calculation and tabulation of the Hall coefficient (R_H) and the conductivity (σ) at ten degree intervals followed by plotting $\log(1/R_H q)$ vs. $1/T$. Fig. 4 shows a typical plot.

The general expression for the Hall coefficient when electrons and holes are present in comparable numbers is

$$R_H = \frac{1}{q} \frac{p - b^2 n}{(bn + p)^2}$$

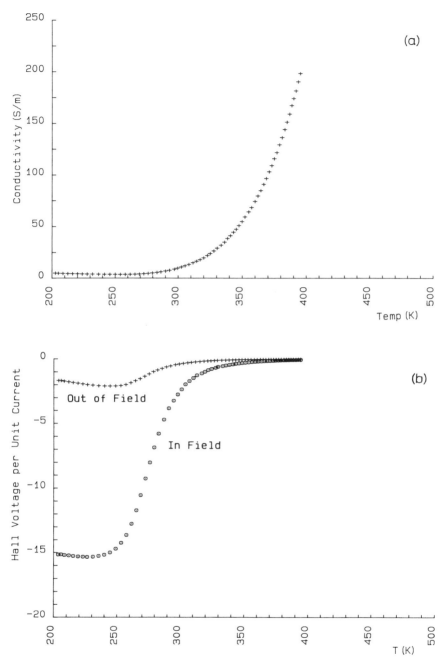

FIG. 3 (a) Computer output plot showing conductivity variation with temperature. During the experiment a second plot showing the low temperature region magnified by 10 is also produced. (b) Computer output showing the Hall + misalignment voltage (in the field) and the misalignment voltage (out of the field).

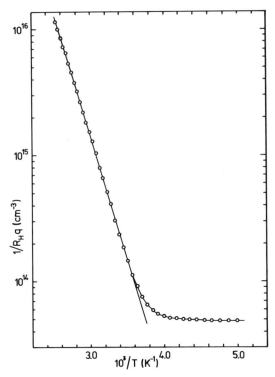

FIG. 4 Arrhenius plot of $1/R_H q$. The slope from the intrinsic region gives a 0K band gap of 0.78 eV. The net ionized donor concentration is seen to be 5×10^{13} cm^{-3}.

where n and p are the free electron and hole concentrations and b is the ratio of the electron mobility (μ_e) to the hole mobility (μ_h). For the purpose of this experiment no distinction is made between the Hall mobility and the conductivity mobility. This expression is quoted in the text and students are asked to determine the form of R_H for extrinsic n-type material ($R_H = -1/nq$), extrinsic p-type material ($R_H = 1/pq$) and intrinsic material ($R_H = (1-b)/n_i q(1+b)$).

Using these results they are asked to identify the intrinsic and extrinsic regions on the $\log(1/R_H q)$ vs. $1/T$ plot, to state whether the material is n-type or p-type and to determine the effective dopant concentration. The majority of students have used the negative values of R_H from the data to show the material to be n-type, but a few have also cited the additional evidence that the Hall coefficient does not change sign on heating the sample.

The intrinsic carrier concentration is given by

$$n_i = \sqrt{N_c N_v} \exp\left(-\frac{E_g}{2kT}\right)$$

and students are asked to use this expression to determine the band gap, E_g. Some students have pointed out the small error involved in neglecting variation of the effective densities of state, and in one case a student pointed to the temperature variation of E_g and showed that the slope of the Arrhenius plot gave the value of E_g at absolute zero.

Finally, students are asked to determine the product $R_H\sigma$ in the case of extrinsic n-type material ($R_H\sigma = -\mu_e$), p-type material ($R_H\sigma = \mu_h$) and intrinsic material ($R_H\sigma = (\mu_h - \mu_e)$), and to plot $\log(-R_H\sigma)$ against $\log T$ (Fig. 5).

By extrapolation of the data from the extrinsic region, a value of the electron mobility at 300 K can be found, and by a similar extrapolation from the intrinsic region ($\mu_e - \mu_h$) — and hence μ_h — can also be determined. From the slope of the extrinsic region the students are asked to find the power law relating μ_e to T and to use this value to determine the dominant scattering mechanism. A typical relationship found from the data in $\mu \propto T^{-1.7}$ which compares with a power of -1.5 predicted from the simple theory of lattice scattering.

4 CONCLUSIONS

Experience in the teaching laboratory has shown the experiment to generate a high level of interest. Although this is largely due to the novelty of using a computer-based measurement system rather than the particular topic

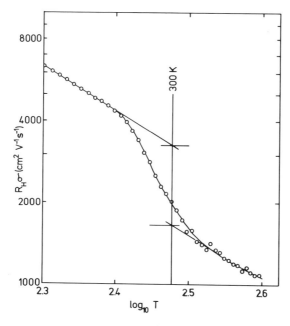

FIG. 5 Log $R_H\sigma$ vs. log T. The slope of the low temperature (extrinsic) region shows $\mu_e \propto T^{-1.6}$. Extrapolation gives a 300 K value of electron mobility as 3300 cm^2 V^{-1} sec^{-1}. Extrapolation of the intrinsic region gives ($\mu_e - \mu_h$) at 300 K as 1700 and thus μ_h as 1600 cm^2 V^{-1} sec^{-1}.

under study, it has become clear that the ease with which high-quality data is obtained encourages a great deal of care in analysing the results. There has been a large amount of feedback from students in the form of questions on specific points concerned with the writing up of the report.

REFERENCE

[1] van der Pauw, L. J., *Philips Tech. Rev.*, **20**, p. 220 (1958).

13

A MICROCOMPUTER-ASSISTED POWER SYSTEM LABORATORY

GILL G. RICHARDS* and PAUL T. HUCKABEE†
*Department of Electrical Engineering, Louisiana State University, Baton Rouge, U.S.A.
†Shell Oil Company, New Orleans, LA, U.S.A.

1 INTRODUCTION

The undergraduate Electrical Engineering curriculum at Louisiana State University has traditionally included two compulsory power-oriented courses and a strong offering of power electives. Student interest in the power area has been maintained by continually updating power course contents to include new developments, particularly in the computer and electronics areas. However, updating the power laboratory has proved more difficult because of the expense of new equipment and the difficulties of inserting new technologies into the instructional laboratory setting. To solve the problem, a new laboratory has been assembled which retains the original rotating equipment, but incorporates digital instrumentation and provides sufficient flexibility to demonstrate modern power systems concepts in such areas as transient stability and control.

2 POWER SYSTEM LABORATORY DESCRIPTION

Since the old motor generator sets were retained, equipment modification of the new laboratory consists only of a redesign of the five laboratory benches and the addition of transducer packages and microprocessor units, which provide all the instrumentation for the laboratory. The power system simulation bench panels were redesigned to limit all interconnections to a single easily accessible patchboard and eliminate much of the time previously wasted in basic hookup. One basic strength of the new laboratory is that each student group can still directly observe the physical behaviour of the machinery, which is located at the end of each bench, while becoming acclimatized to the effectiveness of computer interaction.

High-level signals from the bench are conditioned by transducers housed together in a bench-mounted package. In addition to measuring voltage and current phase angles, the transducers measure three phases of r.m.s. voltage, three phases of r.m.s. current, average power and reactive power, for a total of ten separate measurements. They provide 0–1 mA signals to the A/D converters, which are monitored by analog local indicators mounted on the transducer package. The transducers have been protected from laboratory mishaps with fuses and current transformers, also located in the transducer package. This package is patched into the bench panel between the machine and loads for the measurements required.

As indicated in the schematic diagram, Fig. 1, the transducers supply current-loop signals to the A/D interface. The A/D interface, Z-80, and terminal are mounted together on portable console units, which are located adjacent to the laboratory benches. The standard Z-80 microcomputer system is responsible for coordinating the activities involved in the data acquisition and graphic display. A communications terminal is linked to the microcomputer to allow user control of the system via a keyboard input. The system is

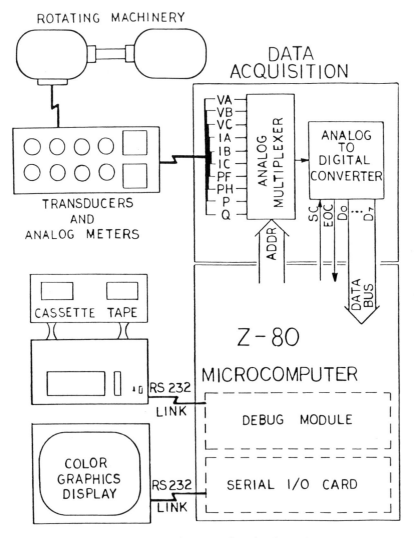

FIG. 1   Laboratory bench schematic.

generally programmed to operate in a conversational mode. The digital part of the computerized laboratory system includes a data acquisition system which multiplexes the analog signals and transforms a selected analog input to an eight-bit digital code, a microprocessor which controls the data acquisition process and generates the display of both the numerical data and graphics, a communications terminal with a CRT graphics display and keyboard for input to the computer, a printer for hard copy output of numerical data (only one for laboratory), and a cassette tape drive for further software development (only one for laboratory).

3 LABORATORY CURRICULUM

An outline of the revised Power Systems Laboratory curriculum is shown in Table 1. The objective in developing these topics was to provide a broader knowledge of modern power system behaviour including modern instrumentation and control concepts. Many parts of these experiments were not possible before the transducers and microprocessors were added. Two of the most convenient new features are the capacity to measure phase angles (particularly in the direct verification of symmetrical components, in experiments 9 and 10) and the availability of the microprocessors (in the off-line mode) which can be programmed, in BASIC, to perform laboratory calculations. This second feature has made it possible to complete all laboratory calculations during the laboratory session, thus eliminating the problems that arise when students first discover their mistaken data as they sit down to perform calculations days after the experiment.

Use of digital equipment with these experiments has required some departure from the usual laboratory procedure. For example, our laboratory generators tend to generate harmonics under the unbalanced conditions of experiments 9 and 10. Because the phase angle transducers for current and voltage that are required here use a zero-crossing criterion, harmonics must be substantially reduced to prevent an erroneous reading. For this reason, these experiments are performed either at a reduced generator voltage, or by using an external voltage source with suitable series impedance.

Feedback control of voltage will be implemented for the final experiments by using the transducers and microprocessors in the on-line mode and incorporating field current control. This addition, which is still in the planning phase, will emphasize the application of control concepts in power, using microprocessor software to simulate exciter response.

4 CONCLUSIONS

The new instructional microcomputer-assisted power systems laboratory has extended the scope of the traditional power laboratory at Louisiana State University. Additional benefits have been increased student involvement with microcomputers, graphics display capability, increased computational capability, and a laboratory experience more closely resembling modern system practice.

TABLE 1 Experiments for new power system laboratory

1. *Familiarization — Power components I*
   Explain laboratory equipment. Examine voltage relationships in various transformer connections. Examine transformer characteristics.
2. *Power components II*
   Examine the behaviour of synchronous machines. Open and short circuit test for saturated reactance. Slip test for x_d and x_q.
3. *Power components III*
   Examine the behaviour of loads: induction machine steady state model, regulation on static loads.
4. *Capacitor as voltage corrector*
   Simulate a regulation condition on laboratory equipment. Insert proper capacitance to improve regulation. Program Z-80 (in BASIC) to supply per-unit reactive power required vs. pf and regulation.
5. *Subtransient reactance*
   Find subtransient reactance by test. Find negative and zero sequence reactance.
6. *Steady state power flow*
   Measure active and reactive power flow directly and by the torque angle equation. Compare with theory. Use phasor display.
7. *Load flow*
   Calculate load flow for the interconnected laboratory system by a separate algorithm. Check results. Observe results of voltage and phase angle pertubations.
8. *Unbalanced operation*
   Calculate voltage and current for three types of unbalanced fault, using the microprocessor and BASIC.
9. *Unbalanced operation*
   Use the laboratory system to simulate the unbalanced operation of experiment 8. Program processor to give positive, negative, and zero sequence components.
10. *Further fault studies*
    Further symmetrical component simulations and calculations.
11. *Swing equation*
    Apply a sudden load and observe shaft dynamics. Find mechanical parameters. Digitally simulate and compare.
12. *Voltage and frequency control*
    Add feedback for voltage control (when available). Calculate and verify stability limit.
13. *Further stability studies*
    Simulation of equal area criterion. Effects of voltage control (when available).

ACKNOWLEDGEMENTS

The laboratory equipment described was partially funded by an instructional equipment grant from the National Science Foundation. The laboratory curriculum was developed under a grant from the Westinghouse Corporation. Microcomputer equipment was donated by MOSTEK.

Part 4

COMPUTER-AIDED DESIGN

14

DEVELOPMENTS IN COMPUTER AIDED CIRCUIT DESIGN

D. BOARDMAN and I. R. IBBITSON
Department of Electrical, Electronic and Control Engineering, Sunderland Polytechnic, England

1 INTRODUCTION

Since the invention of the electronic computer, engineers have been aware of its potential with regard to circuit-analysis. Yet, even now, no readily-available, commercial package meets all the requirements of the circuit design engineer.

The needs of the circuit designer have been analysed in several texts[1,2], and it is clear that their fundamental requirements can be associated with a computer's ability to model, analyse and optimise an initial design. This is shown in Fig. 1, where the conventional and computer-based design procedures are outlined. Both strategies are organised around the design-by-analysis principle. In this situation, the computer's speed is its main advantage, allowing the designer more experimentation time than the conventional approach. However, in the design-by-analysis role, the computer is only simulating the breadboard and test phases of the conventional design procedure, requiring the user to modify the circuit according to the results of an analysis. This, in itself, is useful but does not utilize the computer's full potential in the design procedure. This can be achieved when automatic modification is performed using optimisation and tolerancing routines.

The modelling stage replaces the breadboard but not the initial design, thus making it simply a construction period in which circuit information is fed into the computer. This is usually achieved using techniques such as branch-node listings or wiring operators. These techniques are illustrated in Fig. 2, and it can be seen that, for a reasonably large circuit, they become increasingly unwieldy. They have been used because they are easily related to nodal and two-port network analysis techniques. Many programs also use special languages for circuit input, which are totally alien to the circuit designer, making him resist the use of Computer-Aided Circuit Design (CACD) packages[3].

The analysis phase involves the computer simulating the circuit and performing a d.c. analysis (only in the case of active circuits), linear a.c. analysis and nonlinear transient analysis. From these analyses, information on the bias conditions, frequency and transient responses and sensitivity of a circuit can be obtained. This information can be represented graphically giving a visual picture of circuit performance, in much the same way as an oscilloscope would, in the conventional test stage.

The final stage in the computer design path has traditionally relied on the

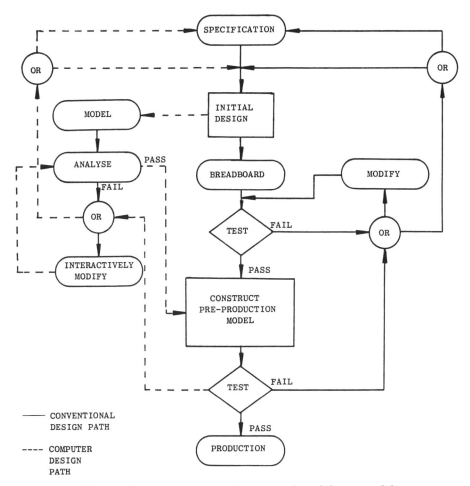

FIG. 1 *The conventional and computer-based design models.*

intuition of the circuit designer. However an optimisation routine in the analyse-interactively modify loop would give a much better indication as to the ultimate potential of the initial design, thus decreasing the time scale in which a pass, redesign or respecify decision can be made.

It is evident that each stage of the computer-based design path is closely related to its conventional counterpart. To replace the conventional design procedure, the CACD package must have the following main attributes;
(a) Fast, economical, computational algorithms,
(b) It must be interactive, user-friendly and graphically orientated.
This, essentially, is the nature of all good Computer-Aided Design and Computer-Aided Learning (CAD/CAL). In a sense, (a) contributes to (b), in that, the quicker the computer responds, the more interactive it is.

Many workers have produced graphs and tables showing different program

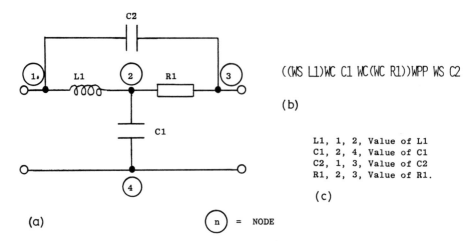

FIG. 2 *(a) Circuit diagram; (b) Wiring-operator circuit description*[12, 13]*; (c) Branch-node circuit description.*

time responses to varying circuits[4, 5]. These are useful in assessing any program with regard to attribute (a), just as 'benchmarks' have their use in comparing different computers. However, the criteria to measure a program's effectiveness, with regard to attribute (b), have not been established. Bowers and Zobrist[4] give the reason for this, when they say that it is an area which is very difficult to 'pin-down' on a graph or chart. Attribute (b) has increased in importance over the years, and is now recognised to be an essential factor in the acceptance of any program by the circuit designer[3, 6, 7, 8].

In the light of the above, it would seem logical to analyse the evolution of CACD and so draw attention to previous mistakes, and attempts to rectify them.

2 A BRIEF HISTORY OF CACD

Until the early 70s many of the available packages, e.g. ECAP, SCEPTRE, CIRCUS, were somewhat unwieldy in size, lacked attractive man-machine interface and required considerable effort on the designer's part to obtain an analysis of all but the smallest circuits. Many of the programs lacked the essential options such as sensitivity, tolerancing and optimisation routines. Due to these imperfections, and enhanced by the euphoria created by the vast amounts of money spent on the subject area[9], a great sense of disappointment ensued when many of these packages became available. This somewhat confirmed the traditional design engineer's wariness of the computer, and helped to maintain a solid resistance against the widespread introduction of CACD, with many simply continuing their breadboard design methods, completely disillusioned. Other factors which also inhibited the use of packages were the early cost of mainframe computers, in terms of initial capital outlay and run time, and the input/output complexities of using such machines. It is recognised that the introduction of mini-computers and time-sharing systems did alleviate this problem to some degree.

Many of the problems were understood by the package programmer and designer, resulting in second-generation programs incorporating more advanced numerical techniques (e.g. using implicit methods of integration to overcome the so-called 'minimum time constant' problem[10,11]), giving improved stability and shorter run times. They also included, for the first time, sensitivity, tolerancing and optimisation routines. Programs such as SUPER*SCEPTRE and SPICE11 seem to sum up the diversification of developments of the early 1970s. Bowers and Zobrist[4] use these two programs to give a typical trade-off picture emphasizing the extremes of development. SUPER*SCEPTRE far outshadowed SPICE11 in terms of flexibility, versatility, support, ease of use and documentation but SPICE11's speed in solving a wide range of simple and complex practical problems was said to be truly phenomenal. Although definite improvements had certainly been made, CACD packages were still not in general use. The reasons for this were twofold; second generation CACD packages were inherently limited as they were, in the main, improvements on earlier realisations, and, as the above example indicates, the two main attributes discussed earlier were not fused together in one program.

As a result, much of the late 70s has been devoted to producing third-generation packages with greater flexibility and functional complexity. However, even as improvements were made in these directions, it became increasingly apparent that they were going to be of little benefit if the typical design engineer still resisted the use of CACD because of poor man-machine interface[3,6,7]. This area is still not very well understood, but it obviously plays a significant role in the attitude of the user to a program. Research workers have started to experiment with what they call graphic processor packages[3] which simply interface the circuit analysis package with the user via a graphical medium. The potential of interactive computer graphics as a medium in which to overcome the problems of the man-computer dialogue was hinted at previously with reference to Fig. 2.

A program example of particular importance in this area is Spence and Apperley's[6] MINNIE system. At the man-computer interface an interactive light pen system was developed and designed to explore the complex interaction between hardware, software and the behavioural characteristics of the human designer.

Emphasis was placed on formulating graphical techniques to overcome problems associated with the behavioural characteristics of the designer, such as psychological closure[16,17], computer response time[16,18], and uncertainty of response to control actions[16]. Also incorporated in the package was a degree of flexibility which allowed the user to modify expressions defining normally calculated circuit properties. The light pen implementation had virtually given the system a pen and paper simplicity which enabled the designer to concentrate on the design and not on its computer representation.

At this point it may be interesting to look at several techniques employed to overcome the frustration that can occur if the designer's behavioural characteristics are ignored. One of the most frustrating and off-putting experiences that can occur when running an interactive computer program, is an apparent lack of computer response to a particular control action. This can occur in CACD when, after a lengthy circuit input period, an analysis command is given and the visual display unit (V.D.U.) screen goes blank for several seconds. Spence and Apperley have overcome this problem by representing the progress of the analysis in the form of a count-down clock. Another approach would be that employed in a program called ELCID[19] where a pattern is generated on the V.D.U. screen in the analysis period. These two techniques, as well as a thermometer-type display used by Essex University Circuits Studies Group[20] are shown in Fig. 3.

3 OTHER FACTORS

As we have seen by briefly surveying the history of CACD, many of the problems involved are related to the circuit designer's needs and lack of good CAD techniques such as interactiveness and graphical representation. Other influences which are of a more general nature in relation to the whole area of CAD are those which involve management, economics and emotion[21]. For example, the cost effectiveness of a CACD unit from the point of view of management can be prohibitive. Not only do management have to find the money to purchase an expensive computer system but often they require

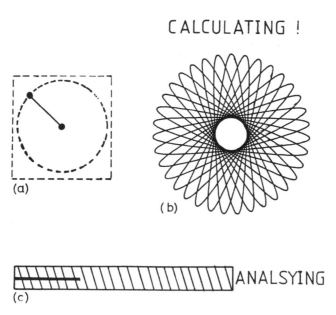

FIG. 3 (not to scale) (a) Count-down clock in Spence and Apperley's MINNIE system; (b) Q.M.C., C.A.T.U.-ELCID; (c) Essex University's C.S.G. thermometer.

additional staff to run the system. On many occasions, unless large volume production is required, an accountant cannot justify the expense.

Emotional problems, although seemingly trivial, can have a considerable effect on the progress of CAD in a company. A frustrating encounter with a computer can be very disheartening, especially to engineers who are not enthusiastic at the thought of learning new techniques, when they feel traditional ones will do equally well. This can make older (usually more senior) men feel very insecure especially when younger colleagues are very enthusiastic and uninhibited at the prospect of using computers. Another area which can cause great concern when CAD is mentioned, is that of unemployment. This, however, seems to stem from ignorance of the computer's present capabilities, rather than any other factors.

These problems are slowly being overcome with the help of time and education. They must be dealt with in the future, as technological advances, economic constraints and competition impose the need for change.

4 MICROPROCESSOR TECHNIQUES

Over recent years, digital CACD has overshadowed its analogue counterpart[22]. This has resulted in the need for increasingly complex analysis packages, to keep pace with circuit component complexity. The problems which have hindered analogue CACD acceptance are amplified on the digital side and are at this point in time a long way from being solved. The complexity involved in designing and analysing VLSI circuits, containing around 100,000 transistors, has created problems which are beyond present CAD techniques. Circuit complexity and the need to improve the productivity of individual designers[23] has meant that semiconductor manufacturers have been forced to provide the stimulus for future CAD development.

It is interesting to note that semiconductor manufacturers have already utilized CAD effectively to produce the microprocessor, a product of integrated circuit technology, which has opened up the newest phase in CACD system development.

Although work has been carried out in various establishments using microcomputers such as the PET[24], the majority of people have not seriously considered the use of microprocessor-based systems in competition with their larger more powerful relations. Yet this is an area which could place CACD within the budget of the smallest firm. Of course initial thoughts would centre on the present day microcomputer limitations such as speed of operation, word length and memory size, but with the advent of the 16-bit microprocessor, dedicated multi-processor CAD systems are becoming a realistic proposition, and have the potential to solve the complex problems, many of which we have outlined, involved in user acceptance.

SPLAP — Sunderland Polytechnic Linear Analysis Program has been developed by the authors to investigate microcomputer graphical man-machine interaction. A Sharp MZ80-K microcomputer is used, though the techniques and algorithms involved can be easily transferred to machines of a

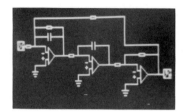

FIG. 4   Drawn circuit diagram.

FIG. 5   Symbol code menu during circuit definition.

similar nature. The main feature of the program, from the circuit designers point of view, is that no knowledge of branch-node listings, nodal admittance matrices or numerical analysis techniques are required, because algorithms have been developed to convert circuit diagrams (see Fig. 4) into numerical analysis information. The circuit designer can develop his ideas on the computer screen in much the same way as he would scribble on a piece of paper. The advantage of drawing it on the computer screen is that the circuit can be analysed immediately, thus overcoming the need to learn any language designed for the computer's benefit. The circuit's position and orientation is controlled by four cursor keys and the circuit components are selected using alphanumeric codes obtained from the symbol code menu at the side of the screen, see Fig. 5. Any additional information required is requested at the bottom of the screen and, once entered, is displayed in the value box at the bottom of the symbol code menu.

SPLAP illustrates that the graphical techniques employed in Spence and Apperley's MINNIE system can be equally applied to low-cost microprocessor-based systems, thus illustrating their potential in overcoming the barriers imposed by the man-machine interface.

ACKNOWLEDGEMENTS

The authors would like to thank Sunderland Polytechnic and NEI Reyrolle Protection for their support in this area of research.

REFERENCES

[1] Calahan, D. A. *Computer-Aided Network Design*, McGraw-Hill, (1972).
[2] Fidler, J. K. and Nightingale, C. *Computer Aided Circuit Design*, Nelson, (1978).
[3] Dyer, J. A., Lawa, A. K., Moran, E. J. and Smart, D. W. 'The use of Graphics Processors for Circuit Design Simulation at GTE AE Labs', ACM, pp. 446–450 (1980).

[4] Bowers, J. C. and Zobrist, G. W. 'A Survey of Computer-Aided-Design and Analysis Programs'. *TR AFAPL-TR*-76-33, WPAFB, Ohio, (April, 1976).
[5] Petrenko, A. I., Timchenko, A. P. and Slyusar, P. B. 'A Comparison of Circuit Design Programs Based on a Set of Test Problems'. IZV. *VUZ Radioelectron* **23**, No. 6, pp. 5–12 (1980).
[6] Spence, R. and Apperley, M. 'The interactive graphic man-computer dialogue in computer-aided circuit design'. *IEEE Trans. Circ. Sys.*, **CAS-24**, No. 2, pp. 49–61 (1977).
[7] Page, W. D. 'Interactive Graphics for Linear Circuit Analysis'. *GTE Automatic Electric Journal*, pp. 233–238 (November 1978).
[8] Agnew, D. 'Scamper — Circuit Design for the 1980s'. *Telesis 1980 Three*, pp. 2–9, (1980).
[9] Neil, T. B. M., 'Techniques for Circuit Analysis (Part Four)'. *Computer Aided Design*, **2**, 3, pp. 38–52, (Autumn 1976).
[10] Branin, F. H. 'Computer Methods of Network Analysis', *Proc. IEEE*, **55**, pp. 1281–1801 (November 1967).
[11] Branin, F. H., Hoesett, G. R., Lunde, R. L. and Kugel, L. E. 'ECAP 11 — An Electronic Circuit Analysis Program'. *IEEE Spectrum*, pp. 14–25 (June 1971).
[12] Adby, P. R. *Applied Circuit Theory, Matrix and Computer Methods*. Ellis Horwood Ltd., (1980).
[13] Penfield, P. *Martha User's Manual*. MIT Press, Cambridge, Mass, (1971).
[14] Spence, R. and Apperley, M. D. 'On the use of interactive graphics in circuit design'. *Digest of papers, in IEEE Int. Symp. on Circuits and Systems*, pp. 558–563 (1974).
[15] Apperley, M. D. 'User-Oriented Interactive Graphics', *DWCUS Europe Conference Proceedings*, Zurich, Switzerland, pp. 273–276 (1974).
[16] Martin, J. *Design of Man-Computer Dialogues*, Prentice-Hall, (1973).
[17] Foley, J. D. and Wallace, V. L. 'The art of natural graphic man-machine conversation', *Proc. IEEE.*, **62**, pp. 462–471 (April 1974).
[18] Miller, R. B. 'Response time in man-computer conversational transactions', *1968 AFIPS Conference Proceedings*, Washington, D.C., Thompson, pp. 462–471 (1968).
[19] *Electronic Circuit Design (ESPC 17E)*, Q.M.C., C.A.T.U.
[20] *ALCAPONE. Linear Analysis Program for Active and Passive Circuits*. Circuit Studies Group, Department of Electrical Engineering Science, Essex University.
[21] Leesley, M. E. 'The uneven acceptance of CAD'. *Computer-Aided Design.* **10**, No. 4, pp. 227–230 (July 1978).
[22] Biancomano, V. 'Logic-simulator programs set pace of computer-aided design'. *Electronics*, pp. 98–101 (October 13, 1977).
[23] Marshall, M. and Waller, L. 'VLSI pushes Super-CAD techniques'. *Electronics*, pp. 73–80 (July 31, 1980).
[24] Fidler, J. K. 'Microcomputer-Aided Circuit Design'. *I.E.E. Colloquium on 'Micro-Computer Aided Design'*, London, England, pp. 6/1–3 (13 May, 1981).

15

LINSIM, A LINEAR ELECTRICAL NETWORK SIMULATION AND OPTIMISATION PROGRAM

L. N. M. EDWARD
Department of Electrical Engineering, University of Canterbury, Christchurch, New Zealand

1 INTRODUCTION

1.1 General description

LINSIM is an interactive real-time linear electrical network analysis, simulation and optimisation program. It is intended to simulate a 'test-bench' and many of the associated a.c. (small-signal) test and adjustment procedures using a conventional network model containing up to 30 nodes and 60 branches.

Nodes are numbered in natural order, beginning with 0 for the reference or ground node.

The user may insert, delete or change any component or node, or command the program to adjust network components so as to maximise or minimise a node voltage or set it to a prescribed value.

Various graphical plots of node voltages (magnitude, decibels or phase) against either frequency or the value of any component, permit very wide-ranging investigation of network performance.

Monte-Carlo simulation, differential and large-change sensitivity analyses and various power calculations are included. Any one of these analyses may be performed at any time by using a structured set of simple, natural commands.

Only a simple visual display unit (VDU) is required, such as a VT52, from which keyboard the network data are entered, manipulated and analyses are commanded.

For the purpose of this paper the term 'manipulate' means any alteration to a network element, whether from the keyboard or under control of an optimising subroutine.

Whilst Computer-Aided-Design (CAD) is now an essential part of network synthesis, its potential for fault diagnosis of existing networks may be less well recognised. LINSIM has, on several occasions, been used to help predict the reason for incorrect network performance by allowing the user to simulate possible failure-modes with the help of its optimisation and parametric plotting capability.

1.2 Origins of LINSIM

LINSIM is a greatly modified and enhanced development of an original program called LINCAD[1], which was highly functional but lacked many refinements which we considered desirable for interactive working. Numerous software additions and changes have transformed the LINCAD of 928 executable statements into LINSIM having over 6000.

1.3 *Network data entry and storage*, (see Table 1)

A network is entered by typing R (for a resistor) or L (for an inductor) and so-on, then following the interactive prompts requesting node-connections and component value. Data may be R, L, C, GM etc., or admittance elements Y, YI, YR, YMR, YMI and YM.

An RLC type of network is a frequency-independent network composed of R, L, C, E, I and GM (voltage-controlled current sources, or VCCS) whereas a Y network is a single-frequency set of complex admittances between the same set of nodes.

An RLC network may be converted to a Y network or vice-versa, but can only be converted properly from Y to RLC when there is no explicit mutual inductance.

Analyses or optimisation may be commanded on either RLC or Y networks.

A network is stored on disc by the command SAVE and it can later be recalled by OLD, so protecting against loss of data in the event of computation failure which causes forced exit.

1.4 *Network structure* (Fig. 1)

Each branch may contain any combination of elements, placed as shown, between nodes J and K. Any node-pair, including J–K, may be chosen as

TABLE 1   *Data types and symbols*

C	Capacitance in μF.
DEG	Phase-angle of E, I, YM or Y. Not valid for initial data entry.
E	Independent voltage-source in volts.
EI	Imaginary part of E.
ER	Real part of E.
F	Frequency in Hz. Default on program entry is 1000 Hz.
FCO	−3dB cut-off frequency of dependent source (VCCS) in MHz. Not valid for initial data entry. Not applicable to YM.
G	Conductance in mhos (Siemens).
GM	Mutual conductance of VCCS in mhos. Cut-off frequency is FCO.
I	Independent current-source in mA.
II	Imaginary part of I.
IR	Real part of I.
K	Coefficient of inductive coupling.
L	Inductance in mH
M	Mutual inductance in mH.
R	Resistance in kΩ.
Y	Admittance magnitude in mhos.
YI	Imaginary part of Y.
YM	Magnitude of mutual admittance of VCCS.
YMI	Imaginary part of YM.
YMR	Real part of YM.
YR	Real part of Y.

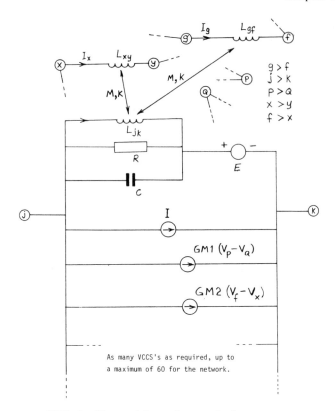

FIG. 1 Network branch canonical structure.

controlling nodes for a VCCS. Only one element of any kind, other than GM or YM, may appear in a branch, but L may couple to as many other inductors as storage permits up to a network total of 36 couplings.

The currents in Fig. 1 are in a positive reference direction when G, J and X are the higher-numbered nodes of the respective branches. If GM2 has controlling node $F<X$, then the reference current in GM2 flows instead from K to J.

Any other branch has the canonical form of branch J–K. A completely filled network could contain 396 separate elements including mutual couplings.

2 THE PROGRAM

2.1 *Command structure*

Good interactive software is tolerant of user-errors and LINSIM performs error-checking and data-validation before continuing with any commanded procedure. The prompt:

Linsim > =

is displayed whenever the program is ready to receive commands or network details.

LINSIM 141

The command-set contains 22 data-types and 101 commands or command-strings. A command takes the form:

COMMAND Qualifier Qualifier Qualifier

where the qualifiers, which are optional, may sometimes be commands or data-types. For example;

TUNE R or HELP CONVERT TO RLC.

An almost unlimited freedom interactively to modify and re-configure a network may be dangerous because a rapid sequence of network changes with intervening analyses, perhaps accompanied by hard-copy, can lead to a rate-of-information-change and sheer mass of data sufficient to disorientate an unwary novice. A designer must learn to hasten slowly, consolidate each change and thoroughly understand its effects before making the next move. To aid comprehension, full details of changes are presented on the VDU as they occur. Several commands will now be described.

2.1.1 *KILL* This command is used to delete a node from the network. The program asks for the node number and deletes all details of components connected to that node.

If a VCCS has the node as one of its controlling-nodes then the VCCS is removed and if an inductor connected to that node has mutual coupling to another inductor, the associated mutual inductance is deleted.

Next, all nodes numbered higher than the deleted node have their numbers decreased by one integer. Details of deleted components and a table of old and new node numbers are then displayed.

2.1.2 *RENUM (renumber)* Frequently a node number needs changing, e.g. so as to keep an orderly set with numbers increasing (say) from left-to-right across the diagram. Using RENUM, any unused integer, other than 0 and not greater than 30, may be assigned to any node. The 'old' node number is then free to be assigned elsewhere. The program checks number availability, perhaps displaying error messages. If all is well it makes the change and then proceeds to make any consequential sign changes to VCCS's or mutual inductances. Then the VDU displays appropriate messages so that the user may update nodes and signs on the network diagram. Thus, though the user may become temporarily confused over the effects, in a complex network, of node renumbering, the program will always keep an accurate and readily-accessible network model which may be examined, at any time, by commanding LIST.

2.1.3 *CONVERT TO Y and CONVERT TO RLC* CONVERT TO Y is used to convert a network, which was entered as R, L, C, etc., to an admittance (Y) network representation at any specified frequency, in which all branches are replaced with complex admittances between the same nodes.

Independent voltage- and current-sources are used unchanged in both RLC and Y networks.

A network, originally entered in admittance form, may be converted to an RLC equivalent by commanding CONVERT TO RLC. Each RLC network branch may then contain R and only L or C depending on the sign of the imaginary part of Y.

If a network branch, as originally entered, contains both L and C and the network is converted to Y, manipulated, and then converted back to RLC, the branch will contain both L and C with the same topography as the original.

Mutual inductance cannot be simply represented in a Y network, but if a network containing mutual inductance is entered in RLC form, converted to Y, and analyses or optimisation, such as TUNE YI indirectly involving mutual inductance, are performed on the Y network, the mutual inductance is properly accounted for because the program utilises both the RLC and Y network forms whilst displaying data or results and accepting commands relevant to a Y network.

The user may cause the Y network performance to be optimised or parametrically plotted explicitly in terms of any inductance, mutually coupled to any other inductance, because the RLC network is automatically converted to a Y network after every RLC network manipulation. Either the RLC or Y network may be worked on at any time.

If manipulations involving M or K are desired and the result is to be in Y network form, these must be done with the RLC network, first making sure that the appropriate frequency is in use, either by using command SHOW F, or directly entering it.

This freedom to move between RLC and Y network forms is very useful when working with high-frequency networks for which active-device data is available in admittance parameters whilst the physical network will employ discrete components.

2.1.4 *TOLER (tolerance)* This is a Monte-Carlo analysis which uses any selection of components with individually specified tolerances.

Either Gaussian or uniform random distributions may be selected for individual elements, reflecting more closely the true situation in practical networks. Distributions are more nearly uniform for components selected from a wide distribution, and Gaussian for those actively tuned or trimmed or in 'as manufactured' state.

A set of up to 500 simulations can be commanded, each one using a different set of perturbations employing mixed Gaussian and uniform distributions.

Comprehensive statistical data are printed concerning any specified node or node-pair for voltage-magnitude, decibels and phase. Some care is needed to interpret statistical results when a mixed distribution is employed, since the concepts of variance and standard deviation are based on Gaussian statistics.

2.1.5 *SENSI (sensitivity)* The most generally efficient component sensitivity analysis may employ the transpose network concept[6]. It is not clear how the general case of mutual inductance can be simply included in this method, so the original LINCAD direct-perturbation algorithm has been kept.

Both differential and large-change sensitivities may be found using this algorithm by changing the perturbation step size from the default-value of 0.1% to any desired value in the range 0.001% to 100%. The lower limit is particularly useful when dealing with a highly frequency-selective network in which even 0.1% may be too large, while the upper limit caters for large-change sensitivity.

The desired perturbation is entered by using the command STEP and following the interactive prompts.

2.1.6 *TUNE* Optimisation, with up to 5 variables specified, employs the well-known Simplex algorithm[5]. Any of the 22 network variables may be manipulated under this command.

A most important network problem relates to the search for conditions leading to oscillation in an active network. All computation is at a frequency F.

By specifying TUNE F and other appropriate variables, such as GM, C, etc., the optimisation may be caused to converge to either a maximum of node voltage, or to a specified very large value, say 1.OE15, yielding from one optimisation both the frequency and threshold conditions.

It sometimes happens that no convergence is possible, and after 100 iterations the program seeks further instructions. To help the user to decide what to reply, four internal variables are displayed at each iteration. As these columns of numbers scroll upwards, their changing patterns suggest what the simplex is doing.

A process which fails to converge will finally cause the scrolling numbers to be unchanging. With experience, the user can often tell whether convergence is likely within another 100 iterations. Without this kind of feedback to reassure him, he will often get impatient with a passive VDU and worry, particularly if the circuit is large and the computer time-sharing load is heavy. There is considerable psychological stress associated with interactive computing and this method of reassuring the user has proven valuable.

2.2 *Documentation* A HELP file, containing over 1000 lines of text, is accessed by prefixing any valid command or command-string or data-type with the word HELP. For example, HELP PLOT LOG causes instructions for obtaining a logarithmic frequency plot to be displayed on the VDU, or HELP BRANCH reveals the canonical structure (Fig. 1), of a network branch.

On initial entry, the program displays instructions explaining how to use the HELP file. If the user types:

PRINT ⟨RETURN⟩

HELP ⟨RETURN⟩

the line-printer will produce a command-list and short introductory commentary.

A User's Manual is available, which comprises the contents of the HELP file, diagrams, additional text and simulated analysis session.

3 MUTUAL INDUCTANCE

Almost all textbooks, apart from those of Guillemin[3] and Chen[4], which present broadly-applicable linear network theory in considerable depth, consistently omit the general case of mutual inductance as it applies to nodal analysis. LINSIM employs an algorithm[8] based on Guillemin's theory.

The network a.c. solution subroutine assembles a nodal admittance matrix from an element-set containing only R, L, C, VCCS, E and I. If working with a Y network it uses the corresponding admittance variables instead. Therefore, mutual inductance can only be included if an equivalent set of inductances, some of which may carry negative signs, is first generated.

Subroutine BIMRIM does this by taking a branch inductance matrix representation (BIM) containing the L and M data, describing the network as the user originated it, and converting it to an indefinite reciprocal admittance matrix (RIM) from which are extracted the equivalent inductors to be used in the nodal analysis.

In general, this algorithm replaces each mutual coupling with a lattice-equivalent of four inductors and modifies the original inductor values. If the coupled inductors share a node, only one additional inductor is generated, resulting in a Pi-section equivalent.

For instance, the 60-branch limit will accommodate, if there are no shared nodes, a 5-winding transformer (45 equivalent inductors), two simple 2-winding transformers (12 equivalent inductors) and three isolated inductors.

The reciprocal inductance matrix approach was chosen for two distinct reasons:
(a) because the RIM structure is easily associated with the nodal formulation and
(b) it produces, through the necessary inversion of BIM, an expanded reciprocal inductance coefficient matrix (EXPAND) which, if used with the node voltage solution, will yield all the inductive branch currents.

The remaining component currents may be obtained from the node voltages and the resistance and capacitance values, giving a complete solution for all voltages and currents in the network at the cost of a single admittance matrix solution for each frequency.

Thus, analyses, plots and optimisation could be performed in terms of either voltage, current or (real or reactive) power.

At present LINSIM does not compute currents but further program development will explore this possibility.

4 EXAMPLES

The following examples have been edited to remove redundant and space-wasting printout. Only a representative few of the 'RLC- and Y-NETWORKS ARE EQUIVALENT ...' panels have been left in. All keyboard input has been underlined editorially to distinguish it from program responses.

4.1 Log-on and network entry

Under VAX/VMS one first depresses the RETURN key, and then follows the VDU prompts. Every command is entered by using the RETURN key (Listing 1).

Next, the network of Fig. 2 will be entered. Only a small portion of the input sequence is shown here in Listing 2.

Entry of 0 for FCO is a special code meaning infinite cut-off frequency, i.e., GM is a frequency-independent VCCS.

Because the program default frequency is 1 kHz, the RLC- and Y-networks are initially equivalent at 1 kHz but later they follow the frequency in use. The Y-network is automatically made equivalent to the RLC-network following:
(a) component entry, change or deletion in the RLC-network
(b) completion of an optimisation process (such as TUNE followed by MAXI, MINI or SET) on an RLC network or
(c) recovery of a network from disc, using OLD.

Other commands such as PLOT, TOLER and SENSI do not change the stored network and provoke no update of equivalence. But if frequency has changed, for example when TUNE F or FIND (bandwidth) are used, then equivalence is re-established.

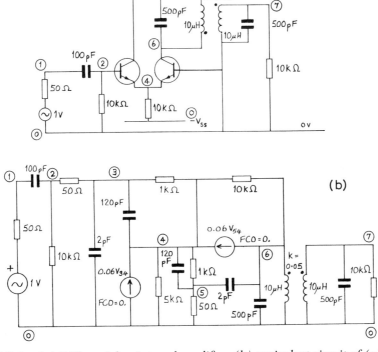

FIG. 2 (a) Differential-pair tuned amplifier, (b) equivalent circuit of (a).

LISTING 1

```
Username: EDWARD
Password:
        Welcome to VAX/VMS Version V2.4

$ R LINSIM
                L  I  N  S  I  M

               FOR ON-LINE ASSISTANCE

          WHEN >= IS THE LAST DISPLAYED SYMBOL

                       Type

                   HELP<RETURN>

*******************************************************
*              N E W    U S E R S                     *
*                                                     *
*   SHOULD OBTAIN HARD-COPY OF THE BRIEFING SHEET     *
*                                                     *
*      By typing PRINT<RETURN> HELP<RETURN>           *
*******************************************************

<RETURN> means the RETURN key. Do NOT type it out!
Linsim >=
HELP
         KEYBOARD COMMANDS AND QUALIFIERS FOR LINSIM
-------------------------------------------------------
ACCURacy          ERROR          MAXI(y)      REAL         TUNE
ADMAT             FIND bw        MINI(y)      RENUMber     VTVM
ALL               HELP           NEW          REPeat       VTVM(y)
ALL(Y)            IMAGInary      NODES        RLC          ::
CLEAR             KILL           NOLISt       SAVE
CLOSE             LIST           NOPRInt      SENSItivity
CONVErt to y      LIST L         OLD          SET
CONVERT TO RLC    LIST(y)        PEAK         SET(Y)
DATA              LOG            PHASE        SHOW
DB                MAGNItude      PLOT         STEP
DEGrees           MAXImise       POWER        TITLE
END               MINImise       PRINT        TOLERance

             DATA-TYPE SYMBOLS
             -----------------
       C        ER      GM      K       Y       YMR
       DEG      F       I       L       YI      YR
       E        FCO     II      M       YM
       EI       G       IR      R       YMI
To get a printout of this type PRINT<RETURN>HELP<RETURN>
Linsim >=
```

4.2 Analyses

Having entered the network, next SAVE it in case a computation failure occurs and LINSIM is thrown out of execution (Listing 4)

At any time a network listing, similar to that following the TOLERance analysis, may be displayed or printed by using the command LIST.

A spot check follows to verify network function, though not necessarily correctness (Listing 5).

Now, suppose we tune the network to 2.25 MHz and then examine the responce characteristics (Listing 6). A frequency-response plot requires specification of the nodes to be plotted, otherwise the program will use the default node-set (1, 2, 3, 4, 5) (Listing 7). The NODES set may be examined at any time (Listing 8). Because the network is narrow-band we use a linear- rather than logarithmic-frequency plot (Fig. 3). Observe that the output mode has been set to dB. Some data on bandwidth, etc., can be obtained using FIND (Listing 9). Sensitivity may be computed for either voltage, decibels or phase. Decibels are shown in Listing 10. Finally, a Monte-Carlo simulation will indicate the likely production-test results (Listing 11). Of course a thorough design investigation would involve many iterations and variants of the above, and other analyses and manipulations.

4.3 Gross network changes

Commands RENUMber and KILL are demonstrated in Listings 12 and 13. They are most commonly used to help correct mistakes and effect major network changes.

Now the session is ended, and we make a graceful exit, remembering to SAVE, if necessary, the final result of the work (Listing 14).

5 CONCLUDING REMARKS

This paper has presented a necessarily sketchy and incomplete description of LINSIM. A true appreciation comes only when one uses the program.

The variety of analyses and the ease with which a network model may be manipulated interactively, result in a convenient and friendly program. Our experience is that students require only about 30 minutes introduction at the terminal to prepare them for serious analysis and design. The average terminal-time used by a second professional year Design class, comprising 3-student groups, to complete the design and characterisation of a typical 15- to 20-node network is about 3 to 4 hours, accumulated in a number of relatively short sessions.

We are able, simultaneously, to run LINSIM from 7 terminals linked to a VAX-11/780 with 1.25 Mb of main memory and obtain quite satisfactory interactive response times, usually less than 1 second.

Although developed primarily for students, LINSIM is used very successfully by professional engineers and is indispensible when analog thick-film networks are to be fabricated in our Hybrid Microelectronics Laboratory[7]. If the demands on fixed memory allocation can be tolerated, then larger networks than are possible within the 30-node 60-branch limit may be handled simply by changing the array declarators and a few other program constants. With continually increasing computer memory sizes, the penalty for temporarily allocating large blocks of 'unused' memory is becoming smaller in real terms, though of course memory should never be squandered.

While programs such as SPICE[2] are very thrifty because of dynamic memory allocation, the fixed arrays of LINSIM greatly simplify both the structures of numerous verification tests and access to components during optimisation.

One would be fortunate indeed if a complicated FORTRAN program, particularly LINSIM which is under continual development, were completely 'bug-free' but our experience so far is quite encouraging.

LISTING 2

```
E
 NODES,VOLTS,DEG
?
1 0 1 0
**************************************************
*    RLC- and Y-NETWORKS ARE EQUIVALENT AT       *
*                                                *
*           1000.000000000000        Hz.         *
**************************************************
Linsim >=
R
 NODES,KILOHMS
?
1 0 50E-3
Linsim >=
C
 NODES,MICROFARAD
?
1 2 100E-6
Linsim >=
L
 NODES,MILLIHENRY
?
7 0 10E-3
Linsim >=
L
 NODES,MILLIHENRY
?
6 0 10E-3
Linsim >=
GM
 NODES,MHOS
?
4 6 .06
   CONTROLLING NODES,FCO IN MHZ
?
3 4 0
Linsim >=
```

LISTING 3

```
K
FIRST NODES,COEFFICIENT OF COUPLING
?
6 0 .05
SECOND NODES
?
7 0
L    6   0     0.10000E-01 MILLIHENRY
  6  0 AND  7  0    M = 0.50000E-03 MILLIHENRY;   K = 0.50000E-01
L    7   0     0.10000E-01 MILLIHENRY
Linsim >=
```

LISTING 4

```
SAVE
ENTER CIRCUIT NAME
DIFFTUNE
DIFFTUNE

Linsim >=
```

LISTING 5

```
ALL
 FREQUENCY IN HZ
?
2.25E6
 DIFFERENTIAL-PAIR DOUBLE-TUNED R.F. AMPLIFIER.

FREQUENCY =     2250000.000000000    HZ

         REAL PART   IMAG PART   MAGNITUDE     PHASE    20*LOG(MAG)
  V  1    0.9904    -0.3651E-02   0.9905      -0.211      -0.083
  V  2    0.9388     0.1315      0.9480        7.974      -0.464
  V  3    0.9339     0.1285      0.9427        7.835      -0.512
  V  4    0.8933     0.1640      0.9082       10.403      -0.836
  V  5    0.8549     0.2037      0.8788       13.401      -1.122
  V  6    1.394     -2.104       2.524       -56.478       8.041
  V  7   -6.028     -6.310       8.727      -133.689      18.817
Linsim >=
```

LISTING 6

```
TUNE C
 NODES OF C
?
6 0
Linsim >=
TUNE C
 NODES OF C
?
7 0
Linsim >=
MAXI
 NODE DIFF TO MAXIMIZE
?
7 0
 FREQUENCY IN HZ
?
2.25E6

C    6   0   0.50791E-03 MICROFARAD

C    7   0   0.50908E-03 MICROFARAD

V  7 - V  0 =   10.123     VOLTS

************************************************
*    RLC- and Y-NETWORKS ARE EQUIVALENT AT      *
*                                               *
*          2250000.000000000     Hz.            *
************************************************
Linsim >=
```

LISTING 7

```
NODES
 NODES TO BE PLOTTED (FILL TO 12 NUMBERS WITH ZEROS)
?
7 6 0 0 0 0 0 0 0 0 0 0
Linsim >=
```

```
PLOT F
INITIAL AND FINAL HERTZ     ,MAX AND MIN DECIBELS   ,NO. OF INTERVALS
?
2E6 2.5E6 20 -5 50
DIFFERENTIAL-PAIR DOUBLE-TUNED R.F. AMPLIFIER.

    * NODE  7   0   . NODE   6   0
                    DECIBELS
            -5.00      0.000E+00  5.00        10.0        15.0        20.0
 0.2000E+07+----------+----*-----+-----------+-----------+-----------+
                           *
                            *
                             *
 HERTZ                        *
                               *
                                *
                                 *
                                  *
                                   *
 0.2100E+07+          +          + *         +           . .+
                                     *
                                      *
                                       *
                                        *
                                         *
                                          *
                                           *
 0.2200E+07+          +          +          .+  *        +
                                             .    *
                                                    *
                                                     *
                                                      *
                                                       *
                                                        *
 0.2300E+07+          +          +          *+          +
                                          *
                                        *
                                      * *
                                    * *
                                  **
                                 *
 0.2400E+07+         +*          +          +           +
                              *
                            *
                          * *
                         -*
                        *
                      *-
 0.2500E+07+          +          .          +           +           +
Linsim >=
```

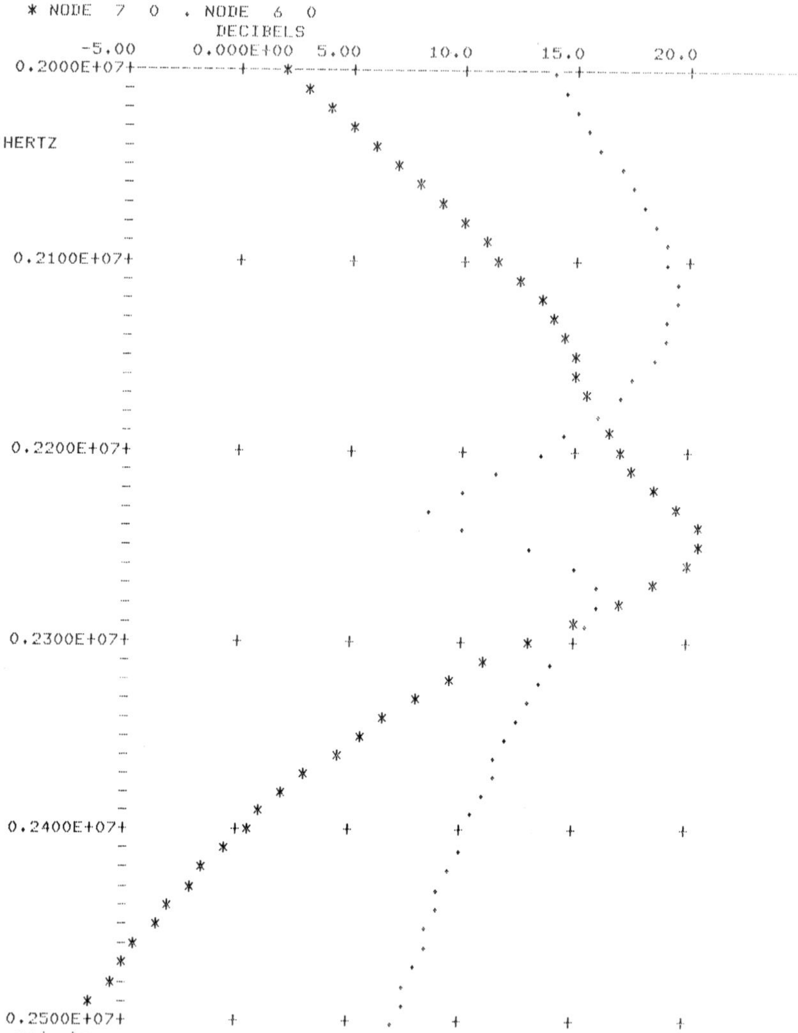

FIG. 3

LISTING 8

```
            SHOW NODES
            CIRCUIT NODES   7  6  0  0  0  0  0  0  0  0  0  0

            NUMBER OF NODES =   7

            Linsim >=
            DB
            Linsim >=
```

LISTING 9

```
FIND
 NODE DIFF TO FIND BW
?
 7 0
 ESTIMATED FREQ IN HZ
?
 2.25E6

CENTRE FREQ =     2249601.745605469      HZ        20.107      DB

F3DB =     2209530.714511871       HZ
F3DB =     2275447.560660839       HZ       65916.84614896774
                                                  = BW IN HZ
Q-FACTOR =           34.13

*************************************************
*    RLC- and Y-NETWORKS ARE EQUIVALENT AT      *
*                                               *
*            2249601.745605469      Hz.         *
*************************************************
Linsim >=
```

LISTING 10

```
SENSI
 FREQUENCY IN HZ
?
 E
DIFFERENTIAL-PAIR DOUBLE-TUNED R.F. AMPLIFIER.

FREQUENCY  =      2249601.745605469      HZ

                           d(DECIBELS  )/DECIBELS
         SENSITIVITY = ------------------------------
                                  dX/X
                                  7- 0        6- 0
  R    1   0   0.5000E-01     -0.434E-02   -0.712E-02
  R    2   0   10.00           0.641E-02    0.105E-01
  R    3   2   0.5000E-01     -0.226E-02   -0.371E-02
  R    4   3   1.000          -0.151E-01   -0.248E-01
  R    5   4   1.000           0.162E-01    0.266E-01
  R    5   0   5.000          -0.421       -0.691
  R    6   4   10.00           0.268E-01    0.440E-01
  R    7   0   10.00           0.215        0.101E-01
  R    4   0   5.000           0.329E-02    0.539E-02
  C    2   1   0.1000E-03      0.700E-02    0.115E-01
  C    3   0   0.2000E-05     -0.824E-02   -0.135E-01
  C    4   3   0.1200E-03     -0.124E-01   -0.204E-01
  C    5   4   0.1200E-03     -0.224       -0.367
  C    6   5   0.2000E-05     -0.185       -0.304
  C    6   0   0.5079E-03     -0.576E-03   -0.945E-03
  C    7   0   0.5091E-03     -0.766E-01    25.3
  L    6   0   0.1000E-01      0.149E-02    0.357
  L    7   0   0.1000E-01     -0.761E-01    25.3
  K    6   0
       7   0   0.5000E-01     -0.843E-02   -0.852
  GM   6   4   0.6000E-01      0.247        0.405
  GM   4   0   0.6000E-01      0.445E-02    0.730E-02


REFERENCE (DECIBELS) =      20.1     12.3
PERTURBATION STEP-SIZE =   0.1000 PERCENT
Linsim >=
```

LISTING 11

```
TOLER
              ENTER DATA TYPE:

      R G L M K C YR YI YM YMR YMI GM FCO E I DEG

      *  TO REVIEW DATA TYPES ALREADY SPECIFIED  *
      *          TYPE:     SHOW                  *
      *                                          *
      *  TO DELETE ALL DATA TYPE SPECIFICATIONS  *
      *          TYPE:     CLEAR                 *
      *                                          *
      *  TO COMMENCE MONTE-CARLO SIMULATION      *
      *          TYPE:     GO                    *
       ?
      C
       COMPONENT NODES
       ?
       6 0
      FOR GAUSSIAN DISTRIBUTION TYPE:  G

      FOR  UNIFORM DISTRIBUTION TYPE:  U
       ?
      G
       PERCENT + or - STANDARD DEVIATION
       ?
      1

              etc.

      R G L M K C YR YI YM YMR YMI GM FCO E I DEG

      *  TO REVIEW DATA TYPES ALREADY SPECIFIED  *
      *          TYPE:     SHOW                  *
      *                                          *
      *  TO DELETE ALL DATA TYPE SPECIFICATIONS  *
      *          TYPE:     CLEAR                 *
      *                                          *
      *  TO COMMENCE MONTE-CARLO SIMULATION      *
      *          TYPE:     GO                    *
       ?
      GM
       COMPONENT NODES
       ?
       4 0
      FOR GAUSSIAN DISTRIBUTION TYPE:  G

      FOR  UNIFORM DISTRIBUTION TYPE:  U
       ?
      G
       PERCENT + or - STANDARD DEVIATION
       ?
      20
       CONTROLLING NODES
       ?
       3 4
              ENTER DATA TYPE:

      R G L M K C YR YI YM YMR YMI GM FCO E I DEG

      *  TO REVIEW DATA TYPES ALREADY SPECIFIED  *
      *          TYPE:     SHOW                  *
      *                                          *
      *  TO DELETE ALL DATA TYPE SPECIFICATIONS  *
      *          TYPE:     CLEAR                 *
      *                                          *
      *  TO COMMENCE MONTE-CARLO SIMULATION      *
      *          TYPE:     GO                    *
       ?
```

LISTING 11 (contd.)

```
GO
FREQUENCY IN HZ
?
F
NODE-PAIR TO BE COMPUTED
?
7 0
NUMBER OF SIMULATIONS
?
100
ENTER A NUMBER TO SEED THE RANDOM NUMBER GENERATOR
?
3656

DIFFTUNE
DIFFERENTIAL-PAIR DOUBLE-TUNED R.F. AMPLIFIER.

FREQUENCY=      2249601.745605469        HZ

MONTE-CARLO ANALYSIS AT NODES    7   0

NUMBER OF SIMULATIONS = 100

ANGLES LIE IN QUADRANTS 0, 2, 3, 0
                             VOLTS          DECIBELS         DEGREES
NOMINAL         =           10.124           20.107          -168.45
MEAN            =            8.2614          18.341          -166.59
MAXIMUM         =           12.701           22.077           177.63
MINIMUM         =            3.6875          11.335          -179.45
STD. DVN.       =            2.1423           2.4753          32.115
VARIANCE        =            4.5895           4.9506         1031.3
3-SIGMA DVN.    =            6.4269           7.4258          96.344


                 COMPONENTS VARIED                          PERTURBATION
C   6   0    0.50791E-03 MICROFARAD                  1.0000     % GAUSSIAN
C   7   0    0.50908E-03 MICROFARAD                  1.0000     % GAUSSIAN
L   6   0    0.10000E-01 MILLIHENRY                  1.0000     % GAUSSIAN
L   7   0    0.10000E-01 MILLIHENRY                  1.0000     % GAUSSIAN
M   6   0    AND   7   0   0.50000E-03 MILLIHENRY   10.000     % GAUSSIAN
K   6   0    AND   7   0   0.50000E-01              10.000     % GAUSSIAN
GM  6   4    0.60000E-01 MHOS                       20.000     % GAUSSIAN
    5   4    CONTROLLING NODES
GM  4   0    0.60000E-01 MHOS                       20.000     % GAUSSIAN
    4   3    CONTROLLING NODES


                 COMPLETE LIST OF COMPONENTS

E   1   0    1.0000        VOLTS       0.00000E+00 DEGREES
R   1   0    0.50000E-01   KILOHMS
R   2   0    10.000        KILOHMS
R   3   2    0.50000E-01   KILOHMS
R   4   3    1.0000        KILOHMS
R   5   4    1.0000        KILOHMS
R   5   0    5.0000        KILOHMS
R   6   4    10.000        KILOHMS
R   7   0    10.000        KILOHMS
R   4   0    5.0000        KILOHMS
C   2   1    0.10000E-03   MICROFARAD
C   3   0    0.20000E-05   MICROFARAD
C   4   3    0.12000E-03   MICROFARAD
C   5   4    0.12000E-03   MICROFARAD
C   6   5    0.20000E-05   MICROFARAD
C   6   0    0.50791E-03   MICROFARAD
C   7   0    0.50908E-03   MICROFARAD
L   6   0    0.10000E-01   MILLIHENRY
L   7   0    0.10000E-01   MILLIHENRY
M   6   0    AND   7   0   0.50000E-03 MILLIHENRY
K   6   0    AND   7   0   0.50000E-01
GM  6   4    0.60000E-01 MHOS         0.00000E+00 FCO IN MHZ
    5   4    CONTROLLING NODES
GM  4   0    0.60000E-01 MHOS         0.00000E+00 FCO IN MHZ
    4   3    CONTROLLING NODES
```

LISTING 12

```
RENUM
WHICH NODE DO YOU WANT TO RENUMBER ?
6
ENTER THE NEW NODE NUMBER
8
Linsim >=
RENUM
WHICH NODE DO YOU WANT TO RENUMBER ?
4
ENTER THE NEW NODE NUMBER
6
                    W A R N I N G

            Sign changes have been made to

            Mutual Inductances, E, I, or VCCS's

                associated with NODE:  6

            Use Command  LIST  and update your

                circuit diagram for NODES:


        8
Linsim >=
```

LISTING 13

```
KILL
ENTER NODE TO BE DELETED
6
DIFFTUNE

DIFFERENTIAL-PAIR DOUBLE-TUNED R.F. AMPLIFIER.

NODE   6 COMPONENTS DELETED ARE:

R      6   4    10.000      KILOHMS
GM     6   4    0.60000E-01 MHOS         0.00000E+00 FCO IN MHZ
    5  4  CONTROLLING NODES
YM     6   4    0.60000E-01 REAL MHOS    0.00000E+00 IMAG MHOS
    5  4  CONTROLLING NODES
Y      6   5    0.00000E+00 +J 0.28269E-04 MHOS
C      6   5    0.20000E-05 MICROFARAD
Y      6   0    0.00000E+00 +J 0.44072E-03 MHOS
C      6   0    0.50788E-03 MICROFARAD
L      6   0    0.10000E-01 MILLIHENRY
M      6   0    AND   7   0   0.50000E-03  MILLIHENRY
K      6   0    AND   7   0   0.50000E-01
Y      7   6    0.00000E+00 -J 0.35463E-03 MHOS

WARNING:  SOME NODES HAVE BEEN RENUMBERED
Amend your circuit diagram as follows:
          OLD No.    NEW No.

             7         6
Linsim >=
```

LISTING 14

```
END
FORTRAN STOP
$ LO
   EDWARD       logged out at 29-MAY-1982 14:19:48.01
```

6 SOFTWARE AVAILABILITY

Copies of LINSIM, in VAX/VMS-readable format on two 8-inch floppy-disks may be obtained, on request, from the author at the cost of the medium plus technician-time, packing and postage. At this time (June, 1982) we are not set up for massive distributions and have no facility for copying to 9-track tape, but this should change within 12 months.

7 REFERENCES

[1] Cornetet, W. H. & Battocletti, F. E., *Electronic Circuits by System and Computer Analysis*, Chaps. 6 and 9 and Appendices, McGraw-Hill (1975).
[2] Nasel, L. W., 'SPICE 2: A Computer Program to Simulate Semiconductor Circuits'. *Memorandum No. ERL-M520*, University of California, Berkeley (1975).
[3] Guillemin, E. A., *Introductory Circuit Theory*. Wiley (1953).
[4] Chen, W. K., *Active Network and Feedback Amplifier Theory*, McGraw-Hill (1980).
[5] Fidler, J. K. & Nightingale, C., *Computer Aided Circuit Design*, Nelson (1978).
[6] Brayton, R. K. & Spence, R., *Sensitivity and Optimisation*, Elsevier (1980).
[7] Edward, L. N. M., 'Microelectronics teaching at the University of Canterbury, New Zealand', *this volume*, p. 59.
[8] Edward, L. N. M., 'BIMRIM: A mutual inductance algorithm for use with nodal analysis programs'. *To be published in the International Journal of Electrical Engineering Education* (1983).

16

COMPUTER-AIDED DESIGN OF DIGITAL SYSTEMS AT FUNCTIONAL LEVEL

L. M. PATNAIK and U. DIXIT
School of Automation, Indian Institute of Science, Bangalore, India

1 INTRODUCTION

Engineering design is a process of decision making in order to produce information to enable the correct manufacture of objects. Thus Computer-Aided Design (C.A.D.) is the use of computer systems to improve decision making, communication and information flow for efficient engineering design. The design process is often classified as follows:

(i) *CAD* In this case, the computer processes information supplied by the designer, yielding results that enable the best decision to be made. This is usually done by simulation, a technique where experimentation is performed on a model of the system.

(ii) *Design automation (D.A.)*: This is an automatic design translation system. The design algorithm is embedded in a program rather than in the mind of the designer.

The electronics industry today holds the place of the fastest growing industry. The circuit designer must meet the ever-growing complexity which only computer-aided tools can enable him to overcome. The overall system design in electronics involves two stages, circuit design and P.C.B. layout design. Both these stages are potential application areas of C.A.D., the former being the more complex and important area of application. Computer simulation is used in the process of circuit design. In addition to providing methods for exact analysis, simulation is cost-effective in reducing design time and eliminating the need for producing expensive circuit prototypes. Simulation of a system involves three major steps[1]:

(i) Modelling of the system.
(ii) Simulation of the system considering the inputs and signal propagation delays.
(iii) Fault location and correction.

By simulation, attempts are made to predict, based on the behaviour of models, the operation of a device or system for a specified set of input conditions. In the design of electrical circuits, generally, two steps are involved:

(i) Simulation of an algorithm describing the circuit function. Here, a system is tested to check for correct outputs, and timing analysis is not required. This design approach is purely combinatorial in nature. This method of attack, known as the functional level approach, assumes that all the timing and synchronization pulses are generated correctly and one need only be concerned with the steady-state aspect of the operation. This approach has been adopted in the present study.

(ii) Simulation of the entire system with signal details where corresponding delays and timing information are to be incorporated.

In this paper the simulation has been carried out at the following two levels for different classes of problems.

(i) *Gate level simulation* — at this level the basic blocks are gates, and the hardware consists of logic circuits comprising gates and their interconnections. The model for a gate is described by a truth table and a Boolean equation relating the outputs to the inputs. The control unit of a hypothetical computer, SAP-1[2], is chosen as an example to demonstrate the design and operation of a control unit in a computer. The interconnection pattern is displayed on a graphics terminal which is a very useful tool in C.A.D. applications.

(ii) *Register level simulation* — at this level, the basic block is the register, and the circuits are described in terms of registers and their interconnections. Portions of arithmetic and logic unit capable of performing signed integer addition, multiplication and division have been simulated. The software package has been implemented in FORTRAN on a PDP-11/35 minicomputer with a VT-11 graphics terminal having a light pen as an input device.

2 SIMULATION OF THE CONTROL UNIT OF A HYPOTHETICAL COMPUTER

The control unit of SAP-1[2], a hypothetical computer used extensively for instructional purposes, has been simulated. A FORTRAN program has been developed to study the function of the control matrix, and to observe various control outputs corresponding to the different inputs from the previous stages. The outputs are checked with the values required for proper execution of the instructions. A graphics routine has been developed to display the control matrix circuit on the VT-11 graphics terminal. The matrix has been made light-pen-sensitive, so that the voltage levels at desired nodes and lines can be observed on the screen.

2.1 *Control unit of SAP*

SAP-1 is a simple, five-instruction, single-address, hypothetical instructional computer. The five instructions are as follows:
(i)   LDA R — load the accumulator with the contents of address R.
(ii) ADD R — add the contents of address R to that of the accumulator.
(iii) SUB R — subtract the contents of address R from that of the accumulator.
(iv) OUT — transfer accumulator contents to the output register.
(v)   HLT — halt the execution of the program.

Each instruction consists of a word of 8 bits; 4 bits are reserved for the mnemonic and 4 bits for the operand address. The SAP-1 machine cycle consists of two parts, the fetch cycle and the execute cycle. The entire machine cycle consists of six timing phases, produced by a ring counter. Phase 1 is called the address phase during which the address in the P.C. (Program Counter) is transferred to the M.A.R. (Memory Address Register). Phase 2 is called the memory phase during which the addressed R.O.M. instruction is transferred from the memory to the instruction register. During phase 3 the P.C. is incremented by one. Phases 4, 5 and 6 form the execution phase.

The control unit (CON) takes as its input a 10-bit word: 4 bits from the instruction register and 6 bits from the ring counter which gives the 6 phases of the machine cycle. The output from the CON is a 12-bit word:

$$CON = C_P\ E_P\ L_M\ E_R\ L_I\ E_I\ L_A\ E_A\ S_U\ E_U\ L_B\ L_O$$

where C_P, E_P etc. are the controlling nodes of the respective registers as shown in Fig. 1. For instance, a high C_P means the P.C. will be incremented, high E_P

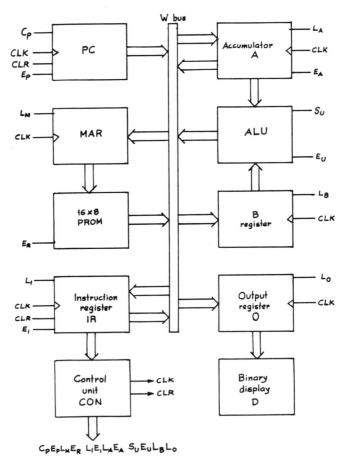

FIG. 1   SAP architecture.

Design of digital systems at functional level 159

and L_M mean that the contents of the P.C. are loaded into M.A.R., etc. During each phase, CON sends out an appropriate 12-bit word which performs the control action. The 12-bit word expected for each phase of the machine cycle is derived as shown below. (i) *Fetch cycle*: In this case, the operations performed in the first three machine phases are common to each of the instructions. During the fetch cycle, CON sends out three control words. In the address phase, E_P and L_M are high, and so the P.C. sets up the M.A.R. via the W bus. A positive clock occurring midway through the address phase loads the M.A.R. with the contents of the P.C. In the memory phase, E_R and L_I are high, thus enabling the addressed R.O.M. to set up the instruction register (I.R.) via the W bus. C_P is the only high control bit during the increment phase. This sets up the P.C., enabling the latter to advance by one at the next positive clock edge. For the ADD and SUB instructions, during the execution cycle, E_I and L_M are high as the instruction field goes to CON and the address field to the M.A.R. in phase 4. In phase 5, the addressed word in the R.O.M. is to be set up in the B-register, and so E_R, L_B are high. In phase 6, the ALU sets up the ACC, and so E_U and L_A are high.

2.2 *The hardware of the control unit*

The control unit shown in Fig. 2 consists of two blocks, the instruction decoder and the control matrix. The instruction decoder accepts a 4-bit input from the instruction field in the I.R. Four lines come out of the decoder, representing LDA, ADD, SUB and OUT instructions respectively. The control matrix inputs are the 6 bits from the ring counter acting as the 6-phase clock, together with the four lines from the instruction decoder. The control matrix output is the 12-bit CON word. The instruction decoder consists of four inverters, and five AND gates. The control matrix consists of 19 AND gates and 6 OR gates,

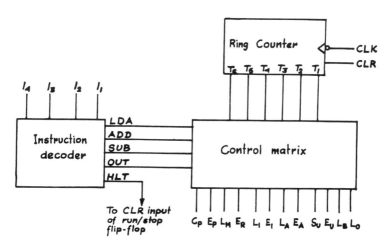

FIG. 2   *Block diagram of SAP control unit.*

driven by the signals mentioned earlier. Lines T_1 to T_6 go high in succession, with only one line high at a time. A 'high' T_1 produces 'high' E_P and L_M bits, a 'high' T_2 produces 'high' E_R and L_I bits and a 'high' T_3 produces a 'high' C_P bit. During the execution phase, T_4 through T_6 go 'high' in succession. At the same time, only one of the decoded inputs can be 'high'. Because of this, the matrix automatically steers 'high' bits to the correct output lines.

2.3 *Simulation details*
The simulation has been performed at the functional level. Hence, each gate has been represented by the Boolean equation relating its output to its inputs. Taking the I.R. bits and the 6-phase clock bits as the inputs, the program delivers the 12-bit CON word as the output for each phase of the machine cycle. This is done for the instructions LDA, ADD, SUB and OUT. The program takes care of both the instruction decoder and the control matrix. For every machine phase, the signal paths corresponding to each output bit are traced. Every gate encountered is replaced by the Boolean relations between the output and inputs. Thus, each output bit is presented by a Boolean expression relating the various I.R. bits and the clock-phase bits.

The program has been organised as follows. The output word in each phase, for every instruction, is obtained by using two nested DO LOOPS. The 4-step outer loop accounts for the simulation being performed for each of the four instructions. The six-step inner loop accounts for the generation of the CON word for the six consecutive machine phases. The I.R. bits are taken as a 4-bit array and the six-phase clock is taken as a 6-bit array. The output word is also an array of 12 bits. The I.R. bits are introduced into the simulation only in the second phase, during which the I.R. is actually loaded. Two subroutines, FALSE and COM are used. The first resets a linear binary array, and the second does bit-by-bit complementation of a linear binary array.

A graphic display device is an important visual aid to the circuit designer. The algorithm developed for simulation can be translated from the Boolean equations to individual gates and their interconnections. The latter can be displayed on the graphics terminal, and this provides the designer with the equivalent of a prototype. The circuit diagram for the control unit of SAP has been displayed on the VT-11 graphics display terminal connected to the PDP-11/35 computer.

3 SIMULATION OF THE ARITHMETIC UNIT
To demonstrate the simulation at register level, signed integer addition, multiplication and division units have been considered. The sign of the operands in all these cases is taken care of by using the two's complement system. An important feature of the simulation at the register level is that all the registers are represented as linear arrays, with array dimensions equal to the register dimensions. Thus, all register transfers are represented in the simulation by array manipulations. The fundamental operation is addition/subtraction. All multiplication and division algorithms involve register

transfers coupled with this operation. Hence, the addition/subtraction algorithm has to be simulated at the logic level because this involves Boolean computation which can be done only at the gate level using Boolean equations.

A FORTRAN subroutine, TADD, simulates the logic circuits of a full adder using the Boolean expressions derived for sum 'S' and carry 'C' from the truth table. It uses a subroutine, TCOM, to convert a negative number to its two's complement form. Subroutines DTLO and OTOL respectively convert operands given in decimal or octal form to binary form. Three arrays represent the two operands and the sum.

3.1 Simulation of the multiplication algorithm

The algorithm for ordinary integer multiplication follows the paper and pencil method[3]. The FORTRAN program MULT, to simulate this multiplier, uses many subroutines to perform rightshifting and leftshifting of registers, two's complementing, ordinary bit-by-bit complementing, two's complement addition and resetting of registers. The sign of the product is taken care of by performing the 'exclusive OR' operation on the MSBs of the two operands, with a 1 leading to a negative product. The multiplication algorithm is applied to the positive number representations of the operands.

Let us consider a powerful direct method for signed number multiplication, the Booth multiplier[3]. It takes two n-bit operands and generates a 2n-bit product and treats both positive and negative numbers uniformly. The algorithm involves the coding of the multiplier bit pattern into a new one. The coded multiplier helps in two ways. First, it enables both negative and positive multipliers to be treated. It also speeds up the multiplication process when the multiplier has a string of ones by avoiding repeated shifting and adding operations. Since the algorithm is discussed in many text books on computer arithmetic/organization, this is not given here. Program BOO simulates the Booth multiplier. Some of the subroutines mentioned earlier are used for this multiplier simulation.

3.2 Simulation of the division algorithm

In the paper and pencil method of dividing integers, if the remainder is negative during the subtraction process, the quotient bit is set to 0, and the dividend is 'restored' by the addition of the divisor to the remainder. Hence the name 'restoring division'. This algorithm can be simulated easily using the subroutines discussed earlier. The 'non-restoring division' method[3] avoids the need for restoring the quantity if a subtraction results in a negative quantity. A program NRDIV simulates this algorithm. In both the algorithms, three points are common. First, both algorithms leave the n-bit quotient in a register Q and the remainder in another register A at the end of the operation. The shifting operation effectively positions the dividend with respect to the divisor, which is equivalent to the reverse process performed in the paper and pencil method. The sign of the result is determined from the sign of the operands, while the algorithms work on the absolute values of the operands.

4 SUBROUTINES USED IN THE SIMULATION

Certain important FORTRAN callable routines have been written to perform secondary operations necessary for the simulation of the control unit of a hypothetical computer and for performing arithmetic functions[4]. These subroutines are as follows.

(i)   SL — This subroutine accepts a linear array representing a register and performs bit-by-bit left shift, and sets the new LSB to 0. It outputs a new array which is the old one left-shifted by one bit.

(ii)   SR — This subroutine shifts a given linear array right by one bit. It sets the MSB of the new array to zero.

(iii) TCOM — This produces an array which is the two's complement of an input binary array.

(iv) COM — This produces an array which is the bit-by-bit complemented form of an input binary array.

(v)   FALSE — This resets a given array to zero. Thus, all bits are set to logical 0.

(vi) TADD — This subroutine performs the two's complement addition of two n-bit linear arrays, producing an n-bit sum and the final carry bit.

(vii) OTOL — This subroutine converts a given octal number into an n-bit binary array.

(viii) LOTO — This subroutine converts the n-bit binary array to its octal form.

(ix) DMLO — This converts a decimal integer to its binary form.

(x)   LODM — This subroutine performs the reverse operation of DMLO.

5 CONCLUSION

The techniques developed in this paper are useful for teaching the concepts of the design of digital systems. Representative examples in the area of design and simulation of the control and arithmetic units of computers have been considered. The use of an interactive graphics terminal helps the designer in interrogating the simulation software package to display the outputs of selected (by a light pen) gates and registers so that fault detection and performance evaluation can be carried out systematically. In addition, one can have a better insight into the operation of the control unit during each phase (T1 through T6 in the example). Design of complicated digital systems involves usage of the subroutines discussed mainly with arrays of larger dimensions. The approach discussed in this paper is extremely useful for teaching the concepts of design/organization of digital computing systems. Details of program listings and implementation aspects can be found elsewhere[5].

REFERENCES

[1] Acken, John M. and Stauffer, Jerry D., 'Logic Circuit Simulation — Part I', *IEEE Trans. on Circuits and Systems*, **1**, No. 1, p. 9 (March 1979).

[2] Malvino, A. P., *Digital Computer Electronics*, Tata McGraw-Hill (1977).

[3] Hamacher, V. C., Vranesic, Z. G. and Zaky, S. G., *Computer Organization*, McGraw-Hill (1978).

[4] Kline, Reymond M., *Digital Computer Design*, Prentice-Hall (1977).
[5] Dixit, Upaindra, 'Computer-aided design of digital systems at functional level', *B.E. Project Report*, Indian Institute of Science, India, (April 1981).

Part 5

AVAILABLE POPULAR SMALL COMPUTERS

17

A REVIEW OF SOME POPULAR SMALL COMPUTERS*

M. J. BOSMAN
Microelectronics Applications Unit, University of Manchester Institute of Science and Technology, England

This review is an attempt to compare the features offered by a number of popular small computers in the hope that anyone thinking of buying one will be able to assess the system best suited to his or her needs, and is based on my own personal views and experience. The machines covered in this review are in no way a complete cross-section of what is on the market. I have, however, tried to select those which, judging by the frequency with which they are mentioned in magazine articles and correspondence, have found widespread use.

The systems reviewed are generally at the lower end of the price range in their simplest form, but the addition of optional peripheral devices and software makes a comparison on the basis of price quite difficult. The basis of the review will be to consider the features of all the machines when offered in their simplest configuration. This generally implies cassette backing store, BASIC as the system language, and a minimum of user memory. System options for upgrading are mentioned as appropriate, as are the input/output facilities provided for the user to improve the usefulness of the machine.

A difficult point in the review has been to indicate 'whose BASIC does what', since the commands and facilities available with the language seem to vary with each manufacturer. Within the limited space allowed by the tabular style of presentation, only a brief indication is given of the language features provided by a given system. A later section of the article adds a little more explanation on what to look for in a particular implementation of BASIC.

Some comments may be in order about the hardware and software resources of a small computer system. To allow a reasonably large program with its associated data to run on the system, at least 8k bytes of RAM should be available. Where a floppy disk backing store and disk operating system are used, more RAM will be required, usually 16k bytes or more. For those computers where a domestic television receiver — as opposed to a monitor — is used for the visual display, the quality of display may be inferior, particularly in the case of colour displays.

---

*This review was compiled in 1980. Though the computers under scrutiny are all still available, prices will have changed (happily, in some cases, these are now even lower).

A variety of graphics types are available on small computers, offering either a set of fixed shapes and graphics characters, or a display of individually-addressable 'picture-points'. The fixed shapes are hearts, diamonds and suchlike, while the graphics characters are patterns within a standard character cell. Each character cell is divided into 3 on its vertical dimension and 2 on its horizontal dimension, giving 6 sub-cells. therefore $2^6 = 64$ pattern combinations of black or white sub-cells form the graphics character set.

A text display area of 40 characters per line, 24 lines, can therefore be considered as composed of 40×2 sub-cells horizontally and 24×3 sub-cells vertically. A 'low resolution' graphics display of 80×72 sub-cells is effectively created and can be used to display simple graphs and diagrams. Better graphics resolution is available when the display can be further sub-divided into individually-addressable picture points ('pixels'). 80 to 250 horizontal picture points give medium-resolution graphics, and high-resolution graphics are obtained with more than 250 horizontal picture points.

The common audio cassette backing store provides a cheap means of storing programs and data. However, it does have disadvantages. Where the cassette recorder is built into the system, data recording is likely to be reliable; the *Commodore Pet* for example uses a redundant recording scheme for reliable data recovery. On the other hand, where the user has to provide the cassette recorder, consistently reliable recordings will more likely be obtained where a better-quality recorder is used. The system should allow programs stored on tape to be accessible by name, to avoid a manual search through the tape. Another important feature is the ability to verify that a program has been transferred from memory to tape without error.

Floppy disk data storage, although more expensive than cassettes, gives rapid storage and retrieval of large programs or 'files'. Two disk drives attached to a system give greater flexibility than a single drive; systems which permit floppy disk attachments may support two, four or more drives. The Disk Operating System software which is supplied with a floppy disk unit should make for a much more satisfactory programming environment. 'File management' operations available might include copying a file from one disk to another (to provide a back-up copy, for example); merging one file with another (to incorporate a routine from a library of 'utilities'), renaming a file; deleting an unwanted file; protecting a file against accidental erasure. The Disk Operating System will also maintain, on the disk, a directory of all files on that disk, which serves as a useful index of programs and their sizes.

Various 'systems' programs will also be supported by the Disk Operating System. These would be typically, a Text Editor program, an Assembler and one or more high-level language compilers (BASIC, FORTRAN, PASCAL, etc.).

SINCLAIR RESEARCH ZX80/ZX81

Price guide: £50/£184

SYSTEM FEATURES

Processor	Z80A
Memory	RAM 1K byte, expandable to 16K bytes. ROM System control and BASIC interpreter — 4K bytes, ZX80; 8K bytes, ZX81 (optional for ZX80).
Backing Store	Cassette interface.
Keyboard	40-key pressure-sensitive keyboard, includes graphics character keys.
Display	Modulated output to TV set. *Text:* 24 lines, 32 characters/line. Standard ASCII character set. *Graphics:* 20 graphics characters, 64h × 44v resolution
Input/Output	Printer output on ZX81 (optional on ZX80). 32 columns, full alphanumerics and graphics. Ability to reproduce display contents.
Resident Operating System	Cassette read/write (programs accessible by name on ZX81); error-checking of statements on entry; input/output management; cursor-controlled screen editing.
Standard Software	4K BASIC: one dimension arrays, 16-bit integer arithmetic, logical operators, syntax checking, compressed program storage. 8K BASIC: floating-point arithmetic, scientific functions, graph plotting, N-dimension arrays.
Optional Software	Educational tape to teach basic mathematics and other programs available.

A review of some popular small computers

ACORN ATOM

Price guide: £120/£385

SYSTEM FEATURES	
Processor	6502
Memory	RAM 2K to 12K on board; external expansion to 40K bytes. ROM System control, assembler, BASIC interpreter — 8K bytes. Expansion to 16K bytes.
Backing Store	Cassette interface (Kansas City standard). 300 bits/sec transfer rate. *Option:* (not yet released) floppy disk drive unit — single drive, 100K bytes capacity, $5\frac{1}{4}''$ disk.
Keyboard	60-key typewriter style.
Display	Video output to monitor, or modulated output to TV set. *Text:* 16 lines, 32 characters/line. Standard ASCII character set. *Graphics:* 64 graphics characters. Optional high-resolution and colour facility (uses 6K bytes RAM); resolution 256h × 192v, black and white; 128h × 192v, 4 colours.
Input/Output	2 × 8-bit input/output ports with handshake lines. Serial input/output port. Optional communication module allows linking to other computers or peripherals (printer, floppy disk, etc.) Internal loudspeaker for tone generation.
Resident Operating System	Cassette read/write and search routine, programs accessible by name; screen editor; program debugging; input/output management (printer drive routine).
Standard Software	Atom BASIC: N-dimension arrays, 32 bit integer arithmetic, logical operators, scientific functions, graph plotting, graphics commands. Assembler for machine-language programming.
Optional Software	9-digit floating-point arithmetic, trig. and hyperbolic functions. Communications software for linking to other computers or peripherals. BASIC extensions for real-time control of laboratory experiments.

NASCOM 2

Price guide: £225/£585

SYSTEM FEATURES	
Processor	Z80A
Memory	RAM 8K bytes; 16K byte expansion board available. ROM 2K byte system monitor; 8K bytes BASIC interpreter.
Backing Store	Cassette interface (Kansas City standard), 300 or 1200 bits/sec transfer rate can be selected. A floppy disk controller card with software, which will support up to three $5\frac{1}{4}''$ disk drives, is available. 160K bytes capacity per drive.
Keyboard	57 key typewriter style. User can change key assignments in software.
Display	Video output to monitor, or modulated output to TV set. *Text:* 16 lines, 48 characters/line, standard ASCII character set, with some additions. *Graphics:* 128 characters; 96h × 48v resolution (plug-in graphic character ROM).
Input/Output	RS-232C/20mA serial port. 2 × 8-bit input/output ports with handshake controls. 77-way bus connector for system expansion.
Resident Operating System	Cassette read/write/verify, programs accessible by name; cursor-controlled screen editor; program debugging; monitor routines accessible to user program; input/output management (supports ASCII terminals connected to serial port).
Standard Software	8K BASIC: (Based on Microsoft 8K BASIC) N-dimension arrays, 7-digit floating-point arithmetic, scientific and logical operations, graphics commands, error messages.
Optional Software	Assembler, Text Editor.

A review of some popular small computers

TANDY TRS-80 Model I

Price guide: £349/£475

SYSTEM FEATURES	
Processor	Z80
Memory	RAM 4K or 16K, expandable to 48K bytes. ROM 4K for Level I BASIC; 12K for Level II BASIC.
Backing Store	Cassette interface. Transfer rate 250 bits/sec with Level I BASIC, 500 bits/sec with Level II. *Option:* up to four $5\frac{1}{4}''$ floppy disk drives may be attached through Expansion Interface. 50K bytes storage on drive 1, 85K bytes on 2, 3, 4. Requires Level II BASIC and 16K bytes minimum system RAM.
Keyboard	53 key typewriter style; 10 key numeric pad optional on 4K versions, standard on 16K versions.
Display	Integral 12" video monitor *Text:* 16 lines, 64 characters/line. Optional selection of double-width characters (32/line) with Level II. Standard ASCII characters. *Graphics:* 128h × 48v. Graphics and text can be combined by software.
Input/Output	40-pin bus connector connects to an optional Expansion Interface which supports peripheral devices and includes parallel port. RS-232C port optional.
Resident Operating	Cassette read/write, input/output management. Cursor controlled screen editor, error trapping, named cassette files and read/write/verify on Level II.
System Standard Software	Level I BASIC: 6-digit floating point arithmetic, one-dimension arrays, graphics. Level II BASIC: N-dimension arrays, user-programmable error messages and error trapping, 16-digit arithmetic, compressed program storage, scientific functions.
Optional Software	Disk BASIC — extended Level II BASIC. Disk Operating System with Utilities and File Commands. Monitor/debug program. Editor/Assembler. Fortran, Cobol, Basic Compiler. Range of educational tapes.

COMMODORE PET

Price guide: £379/£554

SYSTEM FEATURES	
Processor	6502
Memory	RAM 4K/8K/16K/32K bytes ROM Operating System, 4K bytes; Diagnostics, 1K, BASIC interpreter, 8K.
Backing Store	Audio cassette (built-in on 4K and 8K PETs; external for 16K and 32K models). Uses redundant-recording scheme for error correction. *Option:* dual-drive floppy disk unit. Storage capacity 180K bytes per $5\frac{1}{4}''$ disk. Uses IEEE-488 data interface. As an "intelligent" peripheral it uses none of the RAM in the PET.
Keyboard	73 keys, including separate numerical keypad. 16K and 32K PETs have full typewriter size keyboard, 4K and 8K PETs have smaller matrix keyboard.
Display	Integral 9″ black and white video display. *Text:* 25 lines, 40 characters/line. Standard ASCII character set. *Graphics:* 64 graphics characters.
Input/Output	IEEE-488 Interface (allows multiple peripheral connections). 8-bit user port with 2 handshake lines.
Resident Operating System	Cassette read/write/verify, programs accessible by name; cursor-controlled screen editor: input/output management; pseudo-random number generator.
Standard Software	Expanded 8K BASIC: N-dimension arrays, scientific and logical functions, 10-digit floating-point arithmetic, clock function.
Optional Software	Wide range of software packages available, educational, mathematic and scientific. Disk Operating System with File Commands. PASCAL.

SHARP MZ-80K

Price guide: £410/£450

SYSTEM FEATURES	
Processor	Z80
Memory	RAM 20K bytes, expandable to 48K bytes. (BASIC interpreter loaded from cassette into RAM — 14K bytes). ROM 4K bytes system control.
Backing Store	Integral cassette tape unit. 1200 bits/sec transfer rate. Supplied with BASIC interpreter tape. *Option:* dual floppy disk drive. $5\frac{1}{4}''$ disks, storage capacity 140K bytes each.
Keyboard	78 keys: alphabetic (upper and lower case), numeric and graphic symbols. Special editing keys.
Display	Integral 10" black and white monitor. *Text:* 25 lines, 40 characters/line. Standard ASCII character set. *Graphics:* approx. 100 graphics characters available from keyboard.
Input/Output	Input/output supported through add-on Expansion Interface Unit. Allows printer, floppy disk drives, etc. to be connected. Built-in loudspeaker and tone generator.
Resident Operating System	Cassette read/write/verify, programs accessible by name; cursor-controlled screen editor; input/output management; clock function.
Standard Software	BASIC: floating-point arithmetic, N-dimension arrays, scientific functions, graphics.
Optional Software	Editor, Assembler, Symbolic Debugger. BASIC tutorial. Scientific and Educational tapes. Disk Operating System.

APPLE II, APPLE II PLUS

Price guide: £599/£846

SYSTEM FEATURES	
Processor	6502
Memory	RAM 16K byte increments to 48K bytes. ROM System Control, 2K bytes. Applesoft BASIC, 10K bytes/ Integer BASIC, 8K bytes.
Backing Store	Audio cassette interface, 1500 bits/sec. transfer rate. *Option:* Floppy disk subsystem (up to 6 drives). Storage capacity 116K bytes (143K bytes with PASCAL) on $5\frac{1}{4}''$ disks. Requires disk interface card and minimum system memory of 32K bytes.
Keyboard	52 key typewriter style.
Display	Video output for monitor; can use modulator card and TV set. *Text:* 24 lines, 40 characters/line. Standard ASCII character set. *Colour graphics:* 15 colours, 40h × 48v resolution. *High-resolution graphics:* 6 colours, 280 × 192 pixels.
Input/Output	8 board connectors for peripheral device interface cards. 4 analogue (0–150K ohm resistive) inputs. 3 TTL inputs, 4 TTL outputs. Loudspeaker.
Resident Operating System	Cassette read/write; cursor-controlled screen editor; input/output management
Standard Software	Integer BASIC: 16 bit integer arithmetic, N-character names, one-dimension arrays, error check on statement entry. Applesoft BASIC: 9 digit floating point arithmetic, scientific and logical functions, N-dimension arrays, user-programmable error messages and error trapping, high resolution graphics routines.
Optional Software	USCD PASCAL: includes screen-oriented Editor, compiler with extensions for strings, disk files, graphics and system programming, linkage with assembly language routines. Apple FORTRAN. Disk Operating System with Utilities and File Commands.

A review of some popular small computers

INTERTEC DATA SYSTEMS 'SUPERBRAIN'

Price guide: £1650 upwards

SYSTEM FEATURES	
Processor	Two, Z80A (one dedicated to disk operations).
Memory	RAM 32K bytes, expandable to 64K bytes in one 32K byte increment. ROM 2K bytes for system power-up program.
Backing Store	Dual floppy disk drives. $5\frac{1}{4}''$ disks, 350K bytes total storage. Data transfer rate 250K bits/second. Optional double-sided recording gives 700K bytes total storage. Hard disk drives also available.
Keyboard	Typewriter style keyboard plus separate numeric keypad.
Display	Integral 12" video monitor. *Text:* 24 lines, 80 characters/line. Standard ASCII character set. *Graphics:* 64 graphics characters. Optional hardware and software additions for high-resolution graphics.
Input/Output	RS232-C serial port for printer. Additional RS232-C port for linking to host computer or network. Bus connector for system expansion.
Resident Operating System	CP/M Disk Operating System includes disk formatter and file handling utilities, cursor-controlled Text Editor, Assembler, Debugger and input/output management.
Standard Software	(Choose from any of the options) e.g. BASIC — Microsoft 8K BASIC standard/extended/disk version. N-dimension arrays, N-character variable names, logical operators, 7/16 digit floating point arithmetic, user programmable error messages and error trapping.
Optional Software	A wide range of software is available that will run under the CP/M Operating System, e.g. FORTRAN, COBOL, APL, PASCAL, BASIC. Computer communications software.

IMPLEMENTATIONS OF BASIC

The larger proportion of system commands, statements and functions will be common to all the different implementations of BASIC. The notes here are intended to reveal some of the points on which they may differ. The relative importance of any of these features will obviously vary with users' needs and preferences.

(i) *Arithmetic* — the precision with which data values are represented varies widely. Integers may be held as 16-bit or 32-bit, while floating-point numbers vary in representation between 6 and 16 digits.

(ii) *Numerical variables* — some implementations allow only single-character descriptive names, others allow multiple-character names.

(iii) *Array variables* — descriptive names may be restricted as for numerical variables. Some implementations allow only single-dimension arrays, others allow multiple dimensions.
(iv) *String variables* — descriptive names may be restricted as for numerical variables. The number of characters allowed in the string varies from system to system.
(v) *Functions* — the logical functions AND, OR, NOT may or may not be included; of the mathematic/scientific functions, logarithms to base 10, truncate (real to integer) and modulo-arithmetic are available on some implementations only.
(vi) *Error handling* — a valuable facility available on some systems is the provision for user-programmable error messages and error trapping routines. If an error occurs within the program, the program execution goes to a specified line number and continues from there (ON ERROR GO TO ....).
(vii) *Auto line numbering* — some systems provide automatic line numbering of BASIC statements.
(viii) *Graphics commands* — the commands allowed may include individual picture-point addressing, line drawing and colour selection. The graphics facilities of systems vary from minimal to quite comprehensive; anyone particularly interested in a system primarily for graphics capability should examine the facilities very carefully.
(ix) *System monitor* — not all implementations allow the user to leave BASIC and use the facilities of the system monitor for machine code programming.

COMMENT ON PRICE GUIDE FIGURES

As explained earlier, a comparison on the basis of price is quite difficult, but I have selected two levels of system configuration to give some perspective to this exercise.

Each system is available in a minimum form at the lowest price. This means the minimum amount of user memory, use of an external cassette and TV display in some cases, and also, in some cases, a simpler version of BASIC. For the *Acorn Atom*, *Nascom 2*, and *Sinclair ZX80/81* computers, they are obtained in kit form. The price at this level is the first quoted for each computer.

The next level of configuration, and second price quoted, I would regard as being the basis for comparison of systems. The price for each system is with 16k bytes of user RAM, and a black and white video monitor, (I have added £80 to the price for those systems without an integral monitor), with a cassette recorder (I have added £55 to the price of those systems without an integral recorder), and with an 8K byte (or slightly larger) version of BASIC. For those computers available as kits or ready-built, I have taken the ready-built price.

There are some inevitable variations. The *Acorn Atom* accommodates only 12K bytes of RAM on board. The *Sharp MZ80K* provides 22K bytes of user RAM as the nearest size to 16K bytes. The *Sinclair ZX80/81* has no provision for a video monitor, so the price quoted does not allow for one. The *Intertec Data Systems 'Superbrain'* stands apart altogether as it has integral floppy disk storage, an integral monitor and 64K bytes of RAM as standard (the 32K bytes version is only marginally cheaper).

I have chosen the foregoing level of system configuration as giving a useful programming environment in terms of language features and program storage space, and a good-quality display. The price information should therefore be of some value in comparing one system against another when other system features are taken into account.

The cost of expanding a system to include floppy disk drives varies widely. The addition of a single disk drive with disk operating system will cost around £400 to £500; some systems offer a minimum of two disk drives, costing somewhat less than twice the price of a single drive.

(n.b. all prices quoted are nett U.K. prices and exclusive of Value Added Tax. Prices quoted are representative of currently-advertised figures, which do differ from dealer to dealer for the same product.)

ACKNOWLEDGEMENTS

The author would like to thank his colleagues at UMIST for their contributions to the personal review section.

REFERENCES

[1]   *A Guide to the Selection of Microcomputers*, Council for Educational Technology, USPEC 32 (March 1980).
[2]   Li, T. 'Whose BASIC Does What?', *BYTE Magazine* (January 1981)

SOME PERSONAL COMMENTS BY USERS

THE SINCLAIR ZX81

C. S. MILL, Department of Physics, University of Manchester Institute of Science and Technology, England

The *ZX81* was released earlier this year as a successor the *ZX80*. The basic machine sells for about £70 built, £50 in kit form, and as such is about £30 cheaper than its predecessor. It is designed for use with a domestic cassette recorder for programme storage and a UHF television as a monitor. All necessary interconnecting leads are supplied but the power supply is £9 extra if you buy the kit.

The machine used for this assessment was constructed from a kit, which arrived 7 weeks after being ordered. This sort of delay is apparently normal and it is unfortunate that Sinclair's advertisements to date suggest 28-day delivery. A 16K byte RAM extension is sold for about £50 but one was only briefly available for the assessment, so I am only able to say that it worked satisfactorily in the few tests I have been able to conduct. Sinclair is to market a printer for the *ZX81* later this year which is to sell at about £50.

The system is built on a high-quality PCB which is screen-printed and solder-resist coated. Its only real defect is the edge connector which is not gold-plated. Since the circuit only involves four IC's (Z80ACPU, ROM, RAM, and Custom logic IC), a UHF modulator and a sprinkling of discrete components, construction can be completed in about an hour.

I found, initially, that the keyboard was rather difficult to use. It is of the membrane type and although of more or less 'QWERTY' layout it is only about $\frac{2}{3}$ normal size. The problem is made more acute by the single keystroke entry of reserved words, which means that some keys have up to five functions depending upon the mode in which the system is operating. However, once the single keystroke facility has been mastered, entering programmes is quite swift. Those who dislike the small keyboard can replace it with a conventional one (as indeed I have done).

After only a short period of use, the left hand side of the keyboard gets rather hot due to heat dissipation in a voltage regulator, which is located just beneath it. The nominal 9V mains adaptor supplied puts out more like 12V and this cannot be helping the problem. However, it doesn't actually catch fire.

The display gives twenty-four lines of thirty-two characters. The character set is limited to upper case only, in normal and inverse (white on black). Twenty two graphics characters are provided, sixteen of which being those required by the PLOT statement. The bottom two lines of the screen cannot be accessed by PLOT or PRINT statements and, as a result, the screen is 44 pixels high and 64 pixels wide. Resolution is the same in both directions.

The display is not memory-mapped in the usual way as the display moves around in memory. It is possible to find out where it is at any time by peeking at the system variable D FILE (which is thankfully stationary). Characters may thus be poked to the screen with some degree of accuracy. The *ZX81* has two modes of operation as regards the display. In 'fast' mode the screen flickers every time a key is depressed, and goes blank during computation, behaviour which will be familiar to those who have used the *ZX80*. In 'slow' mode the machine maintains a continuous display. This, however is achieved by confining the computation to the blanking periods and reduces the speed by about a factor of four. When 'slow' mode is selected, the top line appears as if in italics, presumably due to the disruption of line sync. during computation. Even in this mode the display is not completely flicker-free as slight twitches occur during the execution of PAUSE and PLOT statements.

The basic machine has an 8K byte ROM containing the basic interpreter and other system software, and 1K byte of RAM. BASIC, as implemented on the *ZX81*, is non-standard in several respects: DATA, READ and RESTORE statements are not available; string manipulation is by LEN and TO statements as MID $, LEFT $, and RIGHT $ are absent; array subscripts start at one and not zero as is normal. Lines may be of indefinite length but can contain only one statement. Unlike the *ZX80*, this machine has floating point arithmetic to $9\frac{1}{2}$ digit precision. All normal mathematical functions are available. The use of a non-standard set of character codes is an unfortunate feature, as it prevents certain standard routines from being used.

Entries from the keyboard undergo a syntax check when 'NEWLINE' is typed. This facility is very good as an inverse 'S' character indicates the location of the error within the line. The system error messages are quite satisfactory. Fifteen different single-character error messages are given, issued together with the number of the line in which the fault appears. The interpretations of these error codes are tabulated in the manual.

In my experience the tape interface performs well. I tested it using an inexpensive portable cassette recorder and have experienced no difficulties at all with it. Program names may be of any length. The display ceases during LOAD and SAVE operations and the screen fills with a series of broad horizontal bands. Although this is not aesthetically pleasing, the pattern is useful in setting up replay levels. There is no VERIFY statement, and no provision has been made for creating or reading data files on tape. Although the latter is somewhat overcome by programmes being SAVEd and LOADed complete with the value of all variables, I feel that it is a major drawback.

Documentation supplied with the machine consists of a neatly-produced 212 page programming book, written by Steven Vickers. (If you buy the kit, a construction sheet is also provided) The book is written for the novice, and as such it is excellent. However, it confines itself almost exclusively to the setting up and programming of the machine in BASIC. For those wishing to extract the most from the system there is much that has been omitted. There is, for example, no explanation of the operating system, nor of system subroutines which could be used. System variables and the allocation of memory is covered in a mere seven pages. Hardware information is limited to the pinout of the 44-way edge connector which carries the data, address and control buses. Since the circuit diagram is given on the instruction sheet, presumably those buying completed machines will not get one.

Despite the lack of information, I have, by trial and error, attached two Z80A PIO's to the *ZX81*, though the port addressing is non-standard, as only addresses for which A$\emptyset$ and A1 are set can be used.

From a teaching viewpoint, I feel that the ZX81 will will find application as a cheap way of providing hands-on experience at an early stage in programming education. In its basic form, with only 1K of RAM the capacity of the machine is easily exceeded (more readily, indeed, than with the *ZX80* because of the extra room taken up by the floating point variables), so that the 16k RAM pack is almost essential. Many of the limitations of the display seem to be associated with trying to get mileage out of the 1K machine. With the rapidly-falling cost of RAM, this may not have been a wise design philosophy.

My main reservation surrounds the lack of detailed documentation. Although most *ZX81* customers will have no call for more information than is provided, the full potential of the system cannot possibly be attained without better documentation. Maybe Sinclair would like to consider selling a supplementary manual as an extra to those who require it.

NASCOM 2

P. J. HICKS, *Department of Electrical Engineering and Electronics, University of Manchester Institute of Science and Technology, England*
C. E. PROCTOR, *Software Sciences Ltd., Macclesfield, England*

The *NASCOM 2* may be purchased either ready-built or as a kit — the subject of this review is one of the earlier kits in which the 8K of on-board static RAM was omitted and replaced instead by a separate board carrying 16K of dynamic RAM.

Assembly of the components onto the densely-packed p.c.b. requires a certain amount of care and probably should not be attempted by the uninitiated. The instructions and diagrams provided in the manual were generally clear and helpful, and the one or two minor errors discovered were easily spotted and therefore of little consequence. These will presumably have been corrected in later issues of the documentation anyway. Unfortunately, however, the video display logic failed to function correctly at first and some probing with an oscilloscope was necessary before a remedy could be applied in the form of a couple of small capacitors. NASCOM, when informed about this, said they were aware of the problem and gave assurances that it has been rectified on all later boards. At this point it is worth mentioning that both NASCOM and the distributor from which the system was purchased (Electrovalue) have always been most helpful in dealing with problems and queries. If required, servicing will be carried out for a fixed charge plus the cost of replacement parts. Setting-up the system consisted of connecting a power supply (NASCOM can supply suitable units) and plugging the output from the UHF modulator into the aerial input socket of a standard TV receiver. Numerous options can be selected using dual-in-line switches on the pcb — these include CPU clock frequency (2 or 4MHz), Memory wait states, Restart address select and control of the serial and tape cassette interfaces. A variety of different cases and racking systems can be purchased for housing the computer pcb and keyboard and these should be used both for convenience and portability and to protect the system. If expansion of the system is envisaged then a rack probably provides the best solution as this can be fitted with a NASBUS backplane. (NASBUS is the NASCOM bus structure.) Into this can be plugged a wide variety of modules to extend the capabilities of the system. Amongst these are memory (RAM, EPROM), I/O, DAC/ADC and colour graphics boards. A mini-floppy disk system is also available.

The Monitor

NAS-SYS 1 is a well-designed operating system. It is built on modular lines from some 50 routines which are invoked via index to a relocatable address table, thus allowing almost unlimited user modifiability and extension to the existing system as well as contributing to a very flexible and powerful monitor. These basic functions form a comprehensive and clearly-documented set to

A review of some popular small computers 181

provide a convenient and permanent basis for program development on the NASCOM 2, since under future versions of NAS-SYS the calling index and functions of these routines will not change. The system is somewhat marred by an unfortunate piece of stack handling in the routine calling mechanism which effectively precludes the use of interrupts in software using these routines, although this problem has apparently been expurgated in an enhanced version of the monitor (NAS-SYS 3) which is now available.

NAS-SYS provides excellent editing facilities with full cursor control, insert/delete features, a scrolling screen and a fixed header line. However, the 48 character wide screen is inconveniently small for high level languages and this is particularly frustrating when one realises that there is a full 16 bytes of unused, invisible margin tagged onto each line. This space appears to have been sacrificed to achieve some simplification of the video RAM hardware, but one would still wish that another solution had been found.

A single-step and breakpoint facility are provided for program debugging. Only one breakpoint is operable at any particular time and this may prove to be inconvenient. At present, the single-step feature is liable to sabotage by the interrupt-phobic monitor (see above). A useful addition to these features would be the option to replace the "automatic register display and return to monitor" after the breakpoint has been executed by a user-modifiable routine which would allow a break on register/store contents.

Machine code patching and memory examination is facilitated by a block copy command and an effective screen edit capability on as much store as can be presented on the screen at one time.

Program storage and retrieval is inconvenient if only because the cassette tape is a poor substitute for a disk. One would have liked the cassette interface to allow programs or data to be restored to different locations in memory to those from which they were copied. It would also have been useful if the monitor had provided for some user-labelling of programs. Since the monitor polls both the cassette tape and the keyboard for input one may visually examine the contents of the tape by simply playing it back. Conversely, the monitor can be directed to channel data destined for the screen to the tape instead. Both ZEAP and BASIC use this method to store selected portions of code.

NASCOM 8K BASIC

NASCOM 8K BASIC is based on Microsoft BASIC, which has become the industry standard, and offers a high degree of compatibility with programs published in books and magazines. In common with other 8 bit BASICs, the NASCOM version is slow. The editor relies almost exclusively on the NAS-SYS screen handling facilities, which while being perfectly acceptable for small programs is inconvenient for large ones. The cassette interface permits storage, examination and retrieval of named files and arrays, although unfortunately the 'name' is only one character long. Only one file is allowed in store at a time and thus the concept of a subroutine library is out of the question. All tape transfers are error checked. The system is well documented and contains some useful additions relating to the *NASCOM*'s screen and graphics facilities. A two byte memory read/write is also included but all memory accesses are made awkward by the inability of BASIC to handle hexadecimal numbers. Programs may be interrupted and current variable contents examined by direct command, thus making debugging easier.

ZEAP Assembler-Editor

ZEAP can be supplied either on cassette tape or programmed on 4 1K EPROMS. Unlike BASIC, ZEAP does provide a subset of standard line editor functions in addition to the screen editing capabilities, but the cassette interface is a bare minimum consisting as it does of listing to or from the tape instead of the screen. The edit buffer is relocatable which allows several programs to be held in store at once, and there is a useful facility for edit buffer salvage following workspace/buffer corruption which is going to happen sooner or later on a machine with unprotected store.

The assembler itself is surprisingly fast and issues clear and informative error messages. It performs all standard Z80 assembler functions (but it has no macro facilities) with the addition of two useful commands which invoke the operating system's relative and subrouting calls. It is worth mentioning here that relocatable code may only be written in Z80 code with the addition of these interpreted calls, since the instruction set otherwise only permits access to the program counter via absolute call, which somewhat defeats the object of the exercise.

The assembler will run programs from specified entry points but contains no facilities for debugging. It is to be hoped that future versions will allow the use of the debugging tools of the monitor with named addresses available in the symbol table.

TANDY TRS Model 1

J. AINSCOUGH, *Microelectronics Applications Unit, University of Manchester Institute of Science and Technology, England*

There were several features which attracted me to the TRS 80 microcomputer system. The first of these was the expandibility of the systems, coupled with a relatively low price of the basic system.

At the bottom of the Tandy price range is a system comprising a CPU, keyboard, 4K BASIC in ROM and 4K of user RAM, plus a UHF modulator. To this must be added the cost of a conventional TV and a cassette recorder. I preferred to begin with a system occupying the next slot up in the tandy range. This consisted of a CPU keyboard, 4K BASIC in ROM and 4K of user RAM, a monitor, and cassette plus separate a.c. power supply. The commissioning of the equipment presented no difficulties, since the vendor-supplied instructions were both clear and concise. The leads, required to connect the individual units together, did provide a rather unwieldy system. However, it does give one the opportunity to arrange the equipment to one's particular needs. In my case, it was possible to operate the system in limited spaces. The 4K BASIC is quite adequate for a beginner, but of limited use to the experienced user, and the editing facilities are also somewhat limited (errors being corrected by retyping the whole line).

The next stage in the evolution of my system was an upgrade to a CPU containing level 2 BASIC and 16K of user RAM. This was quite a simple operation. The original system was delivered to my local Tandy outlet at 9.30 am and the expanded system complete with additional documentation etc, was picked up at approximately 4.30 pm the same day. Again, I found no difficulty in commissioning the system, the improved software being very satisfactory and the improved editing facilities were much appreciated. However, I did note that level 1 and level 2 BASIC are not entirely compatible, the former allowing abbreviated forms of program statements while the latter does not.

With regard to further hardware expansion, Tandy supply an expansion box which incorporates a number of facilities. These include: (i) A memory expansion facility, which allows the user memory to be increased to 48K in increments of 16K, (ii) A printer interface, which supports a range of Tandy printers, (iii) A floppy disk interface which can support up to 4 floppy disks. There is also a range of other peripherals (eg voice input and output system) that are supplied by Tandy. My only reservation with regard to the hardware is that the cases are formed from relatively thin plastics which could pose problems in certain applications.

The software for the system can also be expanded both in terms of the languages and application packages. The former includes both Disk and Cassette-based Z80 assemblers, a Disk-based FORTRAN and a Cassette based TINY PASCAL, (complete with the compiler source written in TINY PASCAL, which could find application in the educational field). The software packages include disk and cassette-based business, education and games packages. I have used both the TINY PASCAL and a Monitor and both have been well documented and have operated satisfactorily.

It would seem that Tandy has intended that it should be the single supplier of this product. However, there are a number of other suppliers of both hardware and software for this product, the latter including the CP/M operating system.

A COMPARATIVE REVIEW OF THE APPLE II AND THE COMMODORE PET

P. F. BRAMELD, Department of Polymer and Fibre Science, University of Manchester Institute of Science and Technology, England

These two machines together have probably made the widest impact on the personal computer market and, having been early entrants to the field, have tended to be compared against each other. Each has its devoted adherents and *Apple* and *Pet* user clubs have flourished. In view of the history of these computers there now exists a substantial body of independent user documentation and applications software, particularly with regard to the *Pet*, but the *Apple* is less well served at present in the UK in this respect. The *Apple*, though, comes with very good supporting documentation in contrast to the *Pet*, whose users might well need recourse to the independent publications.

The *Pet* has a very good screen editor (one of the best around?) which makes the *Apple* editor appear poor by comparison. The *Apple* does however, have a good editor running under the UCSD PASCAL Disk Operating System.

The *Pet* implementation of BASIC is reasonable, but lacks the error trapping function (ON ERROR GO TO ...) and user programmable error messages available on the *Apple* with Applesoft BASIC. The adjustment and control of memory usage by BASIC programs is much simpler on the *Apple*, with commands HIMEM and LOMEM to define the memory space available to a BASIC program. This is, indeed, important on the *Apple*, which has graphics pages interspersed with program spaces. A disadvantage of the *Pet* is the lack of a Reset key to recover from system 'crashes'.

The built-in cassette recorder on the *Pet*, with its redundant recording scheme, gives very satisfactory performance. The cassette routines allow named files and programs, and both data and programs can be saved. The *Apple* cassette system has faster data transfer than the *Pet* but does not allow named files. The user must supply a cassette recorder for use with the *Apple*. The cassette is a reasonably convenient backing store with the *Pet*, but is almost a non-starter on the *Apple*.

The *Apple* must be used with disks and the disk operating system to achieve its full potential. The *Apple* has the useful feature of automatic start up, at power-up, into a program stored on disk. This is not standard on the *Pet*, although it is possible to add a control file on disk to achieve it. The disk unit on the *Pet* is external in that it doesn't use any of the system memory for the disk operating system programs. The *Pet* disk drive is a dual unit, while the *Apple* allows one to six drives to be used.

The outstanding feature of the *Apple* is the graphics performance, and this really needs 48K bytes of RAM and the disk operating system to be fully exploited. The system should also be used with a video monitor for colour graphics work; the modulated TV output is aligned to US standards and gives an inferior display. The system monitor on the *Pet* is very poor, with only a few elementary commands whereas the *Apple* monitor is much better and includes a machine code disassembler program. The *Apple* integer BASIC card has, in addition, a mini-assembler on board.

Peripheral devices are connected to the *Apple* by plugging the appropriate interface card into a slot on the main circuit board (the connectors can be troublesome and exhibit intermittent electrical contact). From a personal point of view this is simpler than the *Pet* system of self-contained peripherals connected by the IEEE interface bus, because the devices are easier to address.

The *Apple* connects to any printer; the *Pet* requires a compatible printer which 'understands' the *Pet*'s non-standard representation of ASCII characters. The *Pet* printer is also very slow: it takes 9 minutes to print a graphics page, whereas a Paper Tiger printer linked to the *Apple* takes only 90 seconds to reproduce a high-resolution graphics page!

From the viewpoint of mainframe computer users coming down to personal computers for the first time, the *Apple* has more resemblance to mainframe practice than the *Pet*. It has a standard keyboard as opposed to the *Pet*'s non-standard one, and uses similar peripheral device addressing techniques.

In conclusion, I feel that the *Pet* has the advantages over the *Apple* when both are cassette-based systems, but the *Apple* is superior with a disk-based operating system and has a particular strength in graphics facilities.

Part 6

COMPUTER-AIDED LEARNING (CAL)

18

EFFICIENT COMPUTER-BASED TEACHING THROUGH TASK SYNDICATION

C. McCORKELL and R. N. WILSON†
*School of Electrical and Electronic Engineering, Ulster Polytechnic (now with the National Institute for Higher Education, Dublin)
†Computer Services, Ulster Polytechnic, Newtownabbey, N. Ireland

INTRODUCTION

In many areas of electrical engineering education including networks, systems dynamics and control, there is a desire to increase computer use. This desire is frustrated in many instances because of the attendant problems of lack of suitable facilities, student inefficiency at the terminal, difficulty of assessment, etc. In the following, we propose a method of instruction which, in conjunction with a properly designed Computer Aided Learning (CAL) package, enables computer use to a greater degree through increased efficiency. Although we concentrated on the teaching of control systems, there is no reason why the approach should not be applied to other areas.

Motivation for computer use in the teaching of control systems arises from:
(1) the possibility of algorithmic solutions to many design problems
(2) the graphical nature of many design techniques
(3) the need to handle repeated computational tasks
(4) the belief that students made familiar with computer use through teaching will use it later in design
(5) the increasing availability of computers

In recent years there have been numerous and varied developments in Computer Aided Design (CAD) facilities for control systems;[1] 'tailored' elements of this research have found use in undergratuate education.[2]

The research and educational objectives are distinct, however, in that in the case of the student designer the aim is the enhancement of skill through choice exercises. For this reason guidance must be incorporated in educational computer packages.

The transformation of a CAD package into a CAL facility involves the inclusion of CAL techniques, such as:
(1) 'help' facilities
(2) specimen answers
(3) error proof input requests
(4) incorporated tutorials

The significance of this is that teaching guidance is now included with syntactic advice on routing within the package.

In accordance with the accepted instructional division of engineering subject matter, the computer is finding use in four broad areas, namely:
(1) assistance with laboratory assignments
(2) solution of tutorial exercises
(3) first-time presentation of material
(4) independent learning

Experience to date, using packages developed at Ulster Polytechnic and including the CAL aids mentioned earlier, has shown that computer use in laboratory and tutorial assignments is straightforward and, in general, efficient.

Attempts to integrate use in a 'first time' presentation were found difficult due to the time required at this stage of the learning process. With the introduction of the computer, the time required threatened to far exceed the time allocated to that particular module. To accommodate computer use, it is apparent that a suitable method of instruction would have to be found.

THE SYNDICATE METHOD[3]

With the syndicate method, a basic statement of a design approach is determined in the classroom. The significant aspects of the technique can be listed and students, in groups, can select an aspect for study.

An illustration of the approach is outlined in Fig. 1. Here, two major phases in the design education process are shown:
(1) a Basic Understanding phase
(2) a Design Competence phase

Following task assignment, student groups undertake the development of their particular section of subject matter, including a computer session, finally making their results available for the following classroom session.

EXAMPLES

Example 1 — Root locus design
A typical schedule for teaching root locus design is given in Fig. 2.

The computer is involved as follows: In (c) and (d) a considerable amount of information is accumulated on transient performance. This is achieved through the series of group tutorials outlined in Fig. 3. The material from each syndicated group is distributed to the class as a whole and a discussion follows to clarify the findings; the tutor can then appraise these findings and present the conclusion that can be substantiated from the evidence.

In the section of the schedule marked (e), the root-locus package[4] is used to permit individual students to generate a number of patterns, some of which are obvious and others not so obvious.

In section (f) of the schedule, the root-locus module is terminated with a design sequence, bringing together the material acquired up to that point.

A typical shared final design exercise would be:[5]
Consider the plant represented by the transfer function:

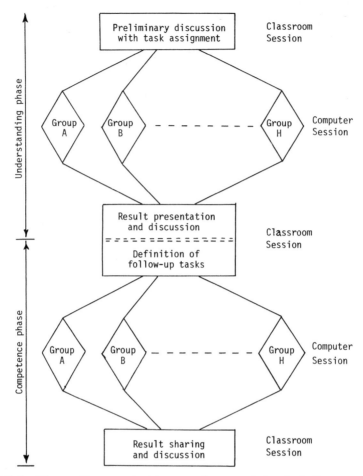

FIG. 1   *The syndicate method in control system design including computer sessions.*

$$G(s) = \frac{2}{s(s+1)(s+5)}$$

Design alternative feedback systems to achieve the following specification:
   (1) steady state error due to a ramp function should be as small as possible
   (2) step response overshoot should not exceed 5 per cent
   (3) settling time should be less than 5 seconds
   (4) rise time should be as small as possible

Loop gain, phase lead and velocity feedback compensators are investigated, the transfer function gain at a point in the s-plane is determined using the root locus package and the resulting compensated model is checked against the specification using the simulation option in the general package.[6]

   This outline of the teaching of the root locus compensation technique

Efficient teaching through task syndication

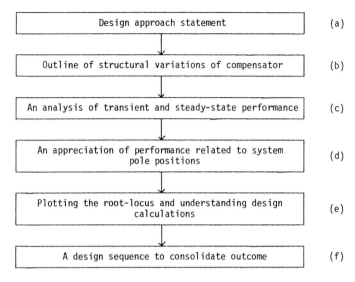

FIG. 2 *Schedule for teaching root locus design.*

demonstrates the use of the computer in conjunction with lectures, and the solution of tutorials.

A separate document has been developed outlining the schedule and including program output which is distributed to students who are studying the topic.

Example 2 — Introducing Nyquist and inverse Nyquist array design techniques
Multivariable frequency response design techniques have been taught in a final year option within a general engineering degree course for the past six years. Latterly the CAD package in use on tutorials has been modified and used in conjunction with the syndicate approach to instruction. The following illustrates the use of the technique.

For a chosen transfer function matrix, the schedule of Fig. 4 is implemented.

Each step occupies computer time and the parallel approach implicit in the syndicate method has been found useful in fully utilising the limited time available. One possible group assignment scheme in this case is shown in Fig. 5. Tasks (a) through (d) involve the Multiple Input/Multiple Output package,[7] whilst (e) and (f) employ the general package for transient response investigations.

DISCUSSION

The syndicated learning approach has been used with class numbers of between eight and twelve students at final year undergraduate level for the past three years. Group members are limited to two students, but experience demonstrates that the students, whilst breaking ground with their partners, subsequently explore further on their own — a consequence of high computer availability.

Group A: Consider the general 2nd order system

$$G(s) = \frac{y(s)}{y_d(s)} = \frac{\omega_n^2}{s^2 + 2B\omega_n s + \omega_n^2}$$

For a particular value of ω_n (say 1 rad/sec) select a range of values for b, falling between 0 and 1, and using the simulation program in PNE1, establish a relationship between 'b' and per cent overshoot for a unit step input.

Group B: Draw a line at 45° from the origin in the S-plane (popularly known as a constant damping line) and select complex conjugate poles corresponding to two distinct positions on the line. For each choice find the system impulse response using PNE1 and comment on the extent to which the system response is damped.

Group C: The following comment is taken from page 6 of your notes:
"For a system to achieve a particular settling time

$$'t'_s - \text{(real part of the poles)} > \frac{4.5}{t_s}$$

determines desired pole positions".
This is a rule of thumb and applies to systems where 'b' is less than 0.8. Use PNE1 to establish the validity of this comment.

Group D: To establish a link between pole positions and rise time consider the effect of varying the real part of the poles from a position close to the imaginary axis to a position some distance away.

Group E: In this final phase of the tutorial you are asked to consider the consequences of adding poles and zeros to an existing pair located close to the imaginary axis. The additional poles should be placed to the left of the existing poles.
The poles closest to the imaginary axis are called dominant poles. Plot transient responses and comment on this "dominance".

FIG. 3 An illustration of task allocation. An appreciation exercise designed to allow the students to explore the link between pole positions and the nature of transient response prior to considering root-locus design.

The packages were initially developed in ICL FORTRAN and ALGOL 60, and made available on the central computer configuration; an ICL 1902T 96K machine, offering access via MAXIMOP to 12 simultaneous users from terminal clusters distributed throughout the campus.

Recently all three packages have been modified to conform to FORTRAN 77 standards and installed on a DEC VAX 11/780 configuration offering access to 24 simultaneous users. This new system is to be expanded to facilitate up to 60 simultaneous users and in the near future, will provide the entire interactive computing service within the polytechnic.

A variant of PNE6, the Multiple Input/Multiple Output package, is also available on a DEC PDP 11/34 configuration within the School of Electrical and Electronic Engineering.

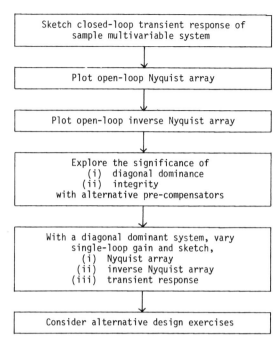

FIG. 4 *Schedule for introducing Nyquist and inverse Nyquist array compensation.*

Group	Task	
A	Compute closed-loop system dynamical equations and plot system step responses.	(a)
B	Plot open loop Nyquist and inverse Nyquist arrays.	(b)
C	Using given pre-compensators compute dynamical equations and plot step responses.	(c)
D	Using given pre-compensators plot Nyquist and inverse Nyquist arrays.	(d)
E	For a given diagonally dominant system vary single loop gain and plot transient responses.	(e)
F	Illustrate the meaning of integrity by producing computed transient responses.	(f)

FIG. 5 *Task allocation in the teaching of multivariable frequency response.*

The original software, as implemented on the ICL 1902T, has been accepted for distribution by ESPE (Engineering Sciences Program Exchange) based at Queen Mary College, London.

CONCLUSION

The computer aided learning approach has been used selectively and introduced with the more traditional methods of teaching. Computer use has considerable motivational value, and, combined with syndication of subject matter, allows parallel activity and, as such, an improvement in instructional efficiency.

REFERENCES

[1] *Computer Aided Control System Design, I.E.E. Conf. Pub. 96*, (1973).
[2] *Control Engineering in Undergraduate Courses, Conference Proc., Teesside Polytechnic* (1976).
[3] Wilkinson, G. M., 'Alternative Methods of Instruction', *Internal Report, Educational Technology Unit, Ulster Polytechnic* (1978).
[4] *PNE3, 'Root Locus Design Package'*. Notes for guidance to students.
[5] Chen, C.T., *Analysis and Synthesis of Linear Control Systems*, Holt, Rhinehart & Winston (1975).
[6] *PNE1, 'Single Input/Single Output Design Package'*. Notes for guidance to students.
[7] *PNE6, 'Multiple Input/Multiple Output Design Package'*. Notes for guidance to students.

19

TEACHING COMPUTER ARCHITECTURAL CONCEPTS USING INTERACTIVE DIGITAL SYSTEM SIMULATION

A. J. WALKER
Department of Electrical Engineering, University of the Witwatersrand, Johannesburg, South Africa

INTRODUCTION

Many educators concerned with undergraduate instruction in digital processes will agree that a primary educational concern is to introduce students to underlying principles which have lasting value and significance. After some years in teaching courses in Digital Processes, from combinational logic design techniques through to micro- and minicomputer applications, it became evident to the author that a significant proportion of graduates still regarded the internal behaviour of computing machines with a sense of mystery. On probing, it became evident that when an existing, functional computer architecture was used as a starting point for discussion and explanation (after an introduction to combinational and sequential logic design techniques with discussion on the functions available in small and medium scale integrated circuits), the following problem areas were among those identified.

- The choice of data multiplexing against bussing arrangements.
- The internal execution of the instruction fetch and execute cycles in relation to the various types of macroinstruction.
- The impact of macroinstruction format on a machines architectural implementation.
- The trade-offs between random logic and a read-only memory based control unit design.
- The trade-offs between vertical and horizontal microinstruction formats.

It was considered that to attempt to deal with the above problems using an existing machine architecture would lead to further confusion, and a more profitable approach would be to start with the architecture of a minimal computer, after the pattern of Heath and Grubb[1] and latterly of Best et al.[2]. The idea of introducing computer architectural design techniques via the minimal instruction computer concept is supported by a number of educators[3,5]. The hardware-implemented minimal instruction computer (MIC) of Heath and Grubb featured a macroinstruction set of four instructions with serial execution of arithmetic. Owing to the absence of any program storage facilities, instructions were serially entered and executed from a front panel switch register. The upgraded version of MIC (or SuperMIC) as implemented by Depledge[2] was designed with a view to removing some of the serious deficiencies of its predecessor. The major objectives in the development of SuperMIC were to illustrate the stored program concept, to provide a visible display of machine states, and an enhanced instruction set. Similar objectives led to the development of the simple computer of Majithia et al.[3].

THE LIMITATION OF HARDWARE-BASED COMPUTER ARCHITECTURE TEACHING AIDS

It is evident from the descriptions given by the above authors that the desired objectives have been achieved. It is, however, the present author's opinion that these hardware-based teaching aids suffer from the same problems which are experienced with vendor-supplied microprocessor learning and evaluation kits. These limitations include the following:

- restricted insight into the execution of macroinstructions at the microinstruction level
- limited access to the internal states of the processor
- architectural inflexibility
- primitive input/output facilities

It is considered that a more useful teaching aid in computer architecture should incorporate the following features.

- It should be possible to observe the state changes with time at the output(s) of any logic element, whether they be for control purposes (bus enables, memory read/write controls, register clock inputs, multiplexer control etc.) data, or addressing purposes (multidigit inputs and outputs to registers, memories, arithmetic/logic units, and the like).
- To provide the highest possible degree of visual comprehension, a visual display must be available with control line state changes displayed in a truth table or waveform format. It is essential to be able to display the output of multidigit devices with a choice of radices.
- The man/machine interface must be enhanced to provide for keyboard input, visual display output, with hard copy facilities to provide a permanent record of activity with the teaching aid.
- The teaching aid must be architecturally flexible to facilitate the natural growth and development of a student's understanding. Once the fundamental behaviour of a stored program machine has been grasped, architectural enhancements can be investigated. The macroinstruction set can be enhanced, memory and register wordlengths extended, and architectural trade-offs examined.

It will be readily acceded that the above requirements may be considered idealistic and desirable, but unlikely to be practical in an undergraduate teaching environment. It will, however, be apparent that if the concept of a hardware-implemented teaching aid is abandoned in favour of an approach using software simulation techniques with an appropriate hardware description language then the above objectives can be achieved. The primary function of the simulator need only be that of logical verification. The success of the simulation approach is heavily dependent on the availability of a suitable software package for the modelling of digital systems.

THE TEACHING REQUIREMENTS OF A DIGITAL SYSTEM SIMULATOR

Commercially-available software packages exist for the modelling of digital systems and their features are described in the literature[6-9]. It is instructive to note, however, that the use of these simulation aids rarely feature in the curricula of universities and colleges. There may be many reasons for their apparent lack of application, but it is considered that one reason may be simply that these packages are more suited to the needs of practicing digital system designers than they are to the requirements of undergraduate instruction. To be effective, the tedium of developing the symbolic network description and data specification must be reduced to a minimum. It is also considered of vital importance that the design iteration time (the time between initiating the execution of a simulation program and observing the results) be reduced to a matter of seconds. At run time, facilities must exist to enable the observation of any output of a logic device in a system and to be able to change the data format of the output of individual logic elements at will. It is not considered a requirement at this level to be concerned about the magnitudes of gate delays and their specifications. This aspect only takes on significance when the digital system under study is logically functional and is required to operate to the limit of the relevant technology. A consideration of technology is relevant to the present exercise, but is considered subordinate to the task of introducing students to computer architectural design philosophies.

It is considered that the following features are required in a simulation tool suited to undergraduate teaching applications.

- It should be capable of modelling networks comprising all the commonly available types of digital integrated circuits, i.e. in the small-scale range (the single binary digit processing elements including the various gate and flip-flop functions), the medium-scale integrated circuit range (the multidigit word processing elements including counters, multiplexers, arithmetic/logic units, decoders, encoders, memories of various types etc.) and large-scale integrated circuits (selected components in a well-known family of bit-slice components).
- It should be capable of modelling networks comprising many hundreds of logic elements.
- It must feature an appropriate hardware description language, in which the effort of developing the symbolic network description and data specification is reduced to a minimum.
- There must be wide flexibility in state and data representation. The outputs of single-digit processing elements must be observable in truth table or waveform format, and a choice of radices should exist to represent the outputs of the multidigit word processing elements.
- It must be structured to minimise the design iteration time. This requires that the system should feature rapid compilation, execution and a high degree of user interaction at runtime.
- It must be possible to verify the logical behaviour of a developed software

model and its physical counterpart. This feature is of significance when the development of software models of highly complex large scale integrated circuits is undertaken, particularly since manufacturer's literature is not always explicit in points of architectural detail which may affect the efficacy of the model.

It was particularly in view of the last-mentioned point that the development of the interactive digital system simulator was not undertaken using the University's time share terminal system, since most vendors of large centralized computer systems strongly discourage the interfacing of special-purpose peripherals to the system. Taking into account the fact that the Department of Electrical Engineering possessed a variety of NOVA minicomputer systems and a software package which facilitated the creation of special-purpose simulation languages, the author undertook to develop the minicomputer-based digital system simulator described in this contribution[10].

AN INTRODUCTION TO THE INTERACTIVE DIGITAL SYSTEM SIMULATOR

The simulation system consists of an executive, and a large number of routines for modelling the behaviour of digital processing elements. The executive provides an environment to support the modelling routines and is largely invisible to the user, except for the presence of a number of executive commands. The activities of compilation of the symbolic network representation statements, runtime data entry and the execution of the simulation are co-ordinated by the executive, but these essential activities are undertaken by the various segments in each modelling routine.

The simulation software consists of beginning and end executive modules, and the user routine libraries comprising utilities, modelling routines of small and medium scale digital integrated circuit devices, and routines for modelling the behaviour of selected components in the Advanced Micro Devices family of bit-slice integrated circuits.

These executive modules and user routines occupy 10K words (16-bits per word) of memory. In a typical single user system with a 32K word memory using the DOS (or RDOS) operating system (occupying up to 10K words), the remaining 12K words are available for the user simulation program workspace. This represents sufficient space for modelling digital networks comprising many hundreds of logic devices. In practice it has been found that 4K words of workspace is sufficient for modelling the logic comprising a microprogrammed 16-bit minicomputer architecture.

Three distinct modes of operation are defined in the executive, namely, the system definition mode, the program mode, and the data entry/runtime mode. The user is made aware of the present executive mode by the particular console prompt associated with that mode. The relationship between the different modes is shown in Fig. 1.

The system definition mode is entered when the simulation system software is loaded. In this mode the user is provided with the potential to model digital

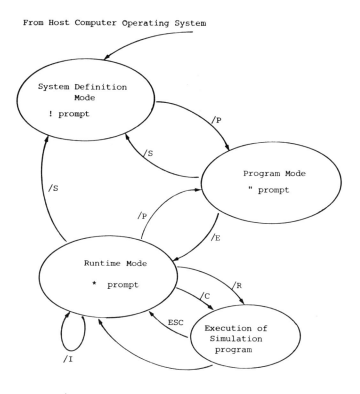

FIG. 1 *Operational relationships between the simulator modes.*

systems consisting of 500 elements of each type. Quite obviously, there are storage requirements associated with each routine, and the actual number of elements that may be specified is dependent on the system memory capacity, the type of modelling elements required, and the size of the network to be modelled. The system definition consists of specifying a modelling element name and the number of elements of that type which are required in the modelled network. The executive command (/P) terminates the system definition mode and the program mode is entered.

In the program mode the statements representing the symbolic network description are entered. Compilation is immediate and a statement is examined once only for compilation purposes. The bulk of a user program consists of the symbolic network description statements. A user program is segmented into three sections, the segments being separated by the keywords, BEGIN, SIMULATE and END. The first section comprises all statements between BEGIN and SIMULATE and are executed once only when the initialization command (/I) is issued. The second segment contains all the statements between SIMULATE and END. The bulk of the network description statements appear here. During compilation, a loop is set up so that these statements can be repeatedly executed. The statements which appear between END and the executive command (/E) are executed once only at the end of a simulation. After the issuing of an (/E) command the data and runtime executive mode is entered. Data which are intended to serve as initialization data for various logic devices are entered under this mode. After the simulator is initialized using the (/I) command, the execution of the simulation program is commenced with the issuing of the (/R) executive command, after which the user must enter the duration of the simulation in the required number of states. An executing program may be interrupted through the use of the keyboard escape key. At the end of the current state, execution is halted. The executive command (/C) allows the user to continue the interrupted simulation exercise.

Any data specification statement and the radix of data representation of any element output may be updated in the runtime mode.

THE HARDWARE DESCRIPTION LANGUAGE FORMAT
The general format of the symbolic network definition and data entry statements is as follows — Name, Number, Parameter 1, ..., Parameter n.

The name is the label associated with the modelling element. The number may be optional parameter, and is used when several elements of the same type appear in the network. For example, a network may contain many gates. If the modelling routine for a logical AND function is called AND, the different gates would be allocated numbers so that one AND gate may be distinguished from another. The statement, AND, 3, would pertain to AND gate three and the input parameters would pertain to that gate.

Parameters 1 to n may be numbers (predominantly in the data statements) or other modelling element routine names (predominantly in the symbolic network description statements).

THE MODELLED COMPUTER ARCHITECTURE AND ITS RELATIONSHIP TO PHYSICAL COMPONENTS
The architecture of a modelled minimal computer is shown in Fig. 2. This minimal computer is used for the sake of example only, and is in no way intended to indicate that the system is restricted to modelling the behaviour of a machine of this complexity only.

The modelled minimal computer features a macroinstruction set of four instructions with memory reference instructions (LOAD and SAVE) and arithmetic instructions (ADD and SUBTRACT)[1]. The microinstruction timing relating to the execution of the macroinstructions is shown in abridged form in Fig. 3. The two most significant bits of the available five-bit memory (and

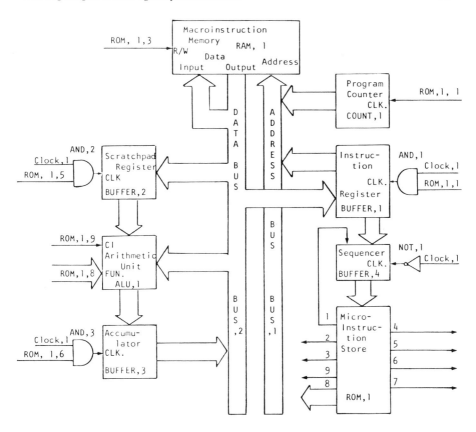

FIG. 2 A simple computer architecture with hardware description language labels.

Instruc-tion	TIME STATE							
	FETCH				EXECUTE			
	L	L→H	H	H→L	L	L→H	H	H→L
LOAD	←	MA:= PC IR:=MD		→	←	MA:= IR SP:=MD		→ PC:=PC+1
SAVE	←	MA:= PC IR:=MD		→	←	MA:= IR MD:=Acc		→ PC:=PC+1
ADD	←	MA:= PC IR:=MD		→		Acc:=Acc+SP		PC:=PC+1
SUBTRACT	←	MA:= PC IR:=MD		→		Acc:=Acc-SP		PC:=PC+1

FIG. 3 Microinstruction timing relationships for the simple computer architecture.

instruction) wordlength constitute the operational code. The remaining three least significant digits are used as the memory address displacement. The machine is limited to the page zero addressing mode. On this account the first eight locations of the macroinstruction memory are conceptually reserved for data storage.

Since no address modification other than that of incrementing is permitted, a simple counter is used as the macroinstruction program counter. The counter increments on the negative edge of the system clock (designated as CLOCK,1) at the commencement of the fetch microinstruction cycle. Four registers (identified as BUFFERs 1 to 4) are used, and each is represented by positive edge-triggered multiple-delay-type flip-flops. Two registers are used in the register-arithmetic unit (the scratchpad register and accumulator). The remaining two are used as the macroinstruction register and macroinstruction memory address register, respectively.

The tristate bus concept is used for the macroinstruction memory data and address busses. The output of each multi-bit device which is to be placed on a bus has associated with it a bus enable control line. In contrast to physical devices in which the tristate concept is integral to the device implementation, the presently-used modelling approach employs the concept of a bus as an independent entity. A word of data to be placed on the bus has associated with it a bus enable input. When the bus enable is asserted, the data word associated with it is placed on the bus.

The eight-word by twelve-bit microinstruction read-only memory is conceptually partitioned into nine independent fields, each of which are designed to control some aspect of the internal resources of the minimal computer. Fields 1 and 2 assert the macroinstruction counter and macroinstruction register address field outputs on the address bus, respectively. Fields 5 and 6 are the scratchpad register and accumulator clock enables, respectively. The arithmetic unit function and carry-in controls are derived from fields 8 and 9, respectively.

The arithmetic unit model represents an implementation of the T^2L device 74181, which is a four-bit wide thirty-two-function arithmetic/logic unit. In the present implementation of the minimal computer only the arithmetic functions of add and subtract are required, which require independent control of the function and carry-in controls.

The statements represented in the symbolic description of the logic components comprising the minimal computer are given in Fig. 4. In most instances a one to one relationship exists between physical devices and the symbolic statements representing those devices.

All comments are preceded by a ♯ character, and these may appear in the context of a simulation program statement. To provide a degree of self-documentation in the simulation, the & symbol is used as a statement continuation character. The responses from the executive are the prompts which may be recognized by the !, " and * characters. The statement ENTER PROGRAM is printed in response to the issuing of the /P command. A double prompt appears in response to the use of a line continuation character. All other characters and statements are entered by the user. Programs are normally stored as disk files, and may be rapidly accessed, compiled and executed.

The wordlength of some elements may be specified by a numerical parameter in the symbolic network representation statement. The arithmetic unit, for example, is specified as being five-bits wide, whereas the physical device, being four-bits wide, would require two such devices in a physical circuit implementation.

The ROM model provides for the specification of up to thirty-two independent fields, in which each field may be independently specified to be one to fifteen-bits wide.

Certain modelling routines (Latch, Buffer and Shift Register models) have been designed to facilitate communication between routines having single and multibit outputs in addition to the primary logical modelling function.

PROGRAM EXECUTION OF THE MINIMAL COMPUTER

The machine code in octal format for the program in the minimal computer which evaluates the expression $D = A + B - C$ is given in the listing of the RAM data statements in Fig. 5. The data for the variables A, B and C are stored in the first three locations in the macroinstruction memory. The macroinstructions for evaluating result D are entered into locations 10_8 to 16_8. The variable, A, is loaded into the scratchpad register, and in the next instruction added to the contents of the accumulator (which has been set to zero at the start of the simulation using the /I command). B is then loaded into the scratch-pad register, and subsequently added to the contents of the accumulator, producing the result $A + B$. C is then loaded into the scratch-pad register and subtracted from the contents of the accumulator, to yield the result $A + B - C$. This result is then stored in location 3 of the macroinstruction memory.

The Print statement is used to display the behaviour of the minimal computer. In Fig. 6 the outputs of the program counter (COUNT, 1), the macroinstruction address displacement (LATCH, 1) and the macroinstruction memory address bus (BUS, 1) are displayed in octal format. The balance of the Print input parameters are displayed in decimal format, which is convenient for checking the arithmetic computations.

It will be readily apparent that all aspects of the machine's behaviour are available for display with the insertion of the appropriate element name and number as a parameter into the parameter list of the Print statement. At runtime the radix and number of digits assigned to output format of any element may be changed to suit the present need. These features have been found to alleviate some of the difficulties normally encountered by students in comprehending the behaviour of complex digital systems.

DISCUSSION

The minicomputer-based interactive digital system simulator described in this contribution has been made available to final-year electrical engineering

FIG. 4   The symbolic network description of the simple computer.

students. A recent digital systems design course extending over a period of fourteen weeks with a class comprising twenty-two final-year students was structured to allow for the design and verification of increasingly complex digital system design exercises. The exercises progressed from an introductory sequential counter design, through the implementation of the add and shift algorithm using microprogramming techniques, to a microprogrammed minimal computer design using bit-slice components, and ending with a

```
"  #
"   BUS    ,2                          &# DATA BUS
"  ...              ,ROM,1   ,4        &# MACROINS. RAM EN.    MICROPROG. ROM FIELD 4
"  ...              ,RAM,1             &# MACROINS. RAM DATA   RAM O/P
"  ...              ,ROM,1   ,7        &# ACCUMULATOR EN.      MICROPROG. ROM FIELD 7
"  ...              ,BUFFER,3          # ACCUMULATOR O/P       BUFFER 3 O/P
"  #
"   AND    ,1                          &# SYSTEM CLOCK 'AND' INSTRUCTION REG. ENABLE
"  ...              ,CLOCK,1           &# SYSTEM CLOCK          CLOCK 1
"  ...              ,ROM,1   ,1        # INS. REG. EN.         MICROPROG. ROM FIELD 1
"  #
"   AND    ,2                          &# SYSTEM CLOCK 'AND' SCRATCHPAD REG. ENABLE
"  ...              ,CLOCK,1           &# SYSTEM CLOCK          CLOCK 1
"  ...              ,ROM,1   ,5        # SCRATCHPAD REG. EN.   MICROPROG. ROM FIELD 5
"  #
"   AND    ,3                          &# SYSTEM CLOCK 'AND' ACCUMULATOR ENABLE
"  ...              ,CLOCK,1           &# SYSTEM CLOCK          CLOCK 1
"  ...              ,ROM,1   ,6        # ACCUMULATOR ENABLE    MICROPROG. ROM FIELD 6
"  #
"   COUNT  ,1                          &# MACROINSTRUCTION PROGRAM COUNTER (5 BITS)
"  ...              ,5                 &# COUNTING DIRECTION    UP (SET HIGH)
"  ...              ,HI
"  ...              ,ROM,1   ,1        # +VE EDGE CLOCK I/P    MICROPROG. ROM FIELD 1
"  #
"   BUFFER ,1       ,WB                &# WORD-IN BITS-OUT CONFIGURED MACROINS. REG.
"  ...              ,AND,1             &# +VE EDGE CLOCK I/P    'AND' 1 O/P
"  ...              ,BUS,2             # DATA WORD I/P         DATA BUS
"  #
"   BUFFER ,2       ,WW                &# WORD-IN WORD-OUT CONFIGURED SCRATCHPAD REG.
"  ...              ,AND,2             &# +VE EDGE CLOCK I/P    'AND' O/P 2
"  ...              ,BUS,2             # DATA WORD I/P         DATA BUS
"  #
"   BUFFER ,3       ,WW                &# WORD-IN WORD-OUT ACCUMULATOR REG.
"  ...              ,AND,3             &# +VE EDGE CLOCK I/P    'AND' 3 O/P
"  ...              ,ALU,1   ,1        # DATA WORD I/P         ALU 1 O/P 1
"  #
"   BUFFER ,4       ,BW                &# BITS-IN WORD-OUT CONFIGURED MICROINS. REG.
"  ...              ,NOT,1             &# +VE EDGE CLOCK I/P    INVERTED SYSTEM CLOCK
"  ...              ,BUFFER,1,5        &# OP. CODE BIT 1        MACROINS. REG. BIT 4
"  ...              ,BUFFER,1,4        &# OP. CODE BIT 0        MACROINS. REG. BIT 3
"  ...              ,ROM,1   ,1        # MICROINS. CLOCK        MICROPROG. ROM FIELD 1
"  #
"   PRINT  ,1                          &# RESULTS DISPLAY IN TRUTH TABLE FORMAT
"  ...              ,CLOCK,1           &# SYSTEM CLOCK          CLOCK 1
"  ...              ,ROM,1   ,1        &# PROGRAM COUNTER CLOCK MICROPROG. ROM FIELD 1
"  ...              ,COUNT,1,1         &# PROGRAM COUNTER       COUNT 1 O/P 1
"  ...              ,AND,1             &# INSTRUCTION REG. CLK  AND 1 O/P
"  ...              ,LATCH,1           &# ADDR. DISPLACEMENT    PSEUDO-LATCH 1
"  ...              ,BUFFER,4          &# MICROINS. COUNTER     BUFFER 4
"  ...              ,ROM,1   ,2        &# DISPL. ADDR. BUS EN.  MICROPROG. ROM FIELD 2
"  ...              ,BUS,1             &# ADDRESS BUS           BUS 1
"  ...              ,ROM,1   ,4        &# RAM DATA BUS ENABLE   MICROPROG. ROM FIELD 4
"  ...              ,ROM,1   ,7        &# ACCUMULATOR BUS EN.   MICROPROG. ROM FIELD 7
"  ...              ,BUS,2             &# DATA BUS              BUS 2
"  ...              ,ROM,1   ,3        &# RAM R/W ENABLE        MICROPROG. ROM FIELD 3
"  ...              ,RAM,1             &# MACROINSTRUCTION RAM  RAM 1 O/P
"  ...              ,AND,2             &# SCRATCHPAD REG. CLK.  AND 2 O/P
"  ...              ,BUFFER,2          &# SCRATCHPAD REGISTER   BUFFER 2 O/P
"  ...              ,ROM,1   ,9        &# CARRY-IN CONTROL      MICROPROG. ROM FIELD 9
"  ...              ,ROM,1   ,8        &# ALU FUNCTION          MICROPROG. ROM FIELD 8
"  ...              ,ALU,1   ,1        &# ARITHMETIC UNIT       ALU 1 O/P 1
"  ...              ,AND,3             &# ACCUMULATOR CLOCK     AND 3 O/P
"  ...              ,BUFFER,3          # ACCUMULATOR           BUFFER 3 O/P
"  #
"   END                                # SIMULATION KEYWORD
"  #
"   /E                                 # END OF SYMBOLIC NETWORK DESCRIPTION
*
*
```

FIG. 4 contd.

microprogrammed computer with sixteen instructions. Some thirty hours of terminal time per student was allocated for the execution of these exercises. An introductory session of two hours is sufficient for most students to develop sufficient expertise to run their own simulation exercises. Each student is expected to acquire a copy of the simulation system manual[10].

The computing resources within the Department include three stand-alone NOVA computing systems and an Eclipse S/140 multiprogramming computer

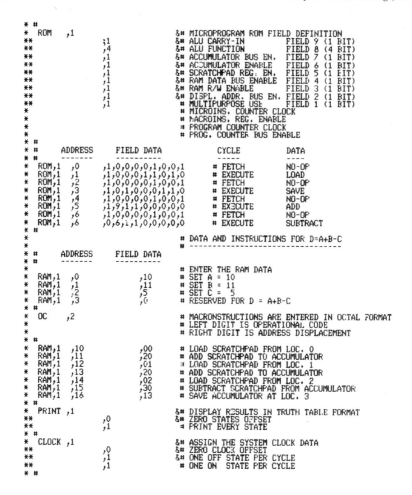

FIG. 5 The runtime and data statements for the simple computer.

system presently supporting four user consoles. The bulk of student simulation exercises are undertaken on this latter system. The present version of simulation software is written in assembler language and is designed to function within the DOS/RDOS and AOS operating systems, as supported by Data General Corporation.

The author is prepared to make the simulator software available to bonafide educators for teaching purposes.

FIG. 6 A simulation of the behaviour of the simple computer.

REFERENCES

[1] Heath, F. J. and Grubb, D. T., 'A digital computer for sixth-forms' *Int. J. Elect. Eng. Educ.*, **3**, No. 1, (1965).

[2] Best, P. J., Depledge, P. G., Escuder, M. A. and Powner, E. T., 'Introduction to microprocessor systems — part 3'. *The Challenge of Microprocessors*, Ed. M. G. Hartley, Manchester University Press (1979).

[3] Majithia, J. C., Banks, W. and Jansen, J., 'Two examples in minicomputer design', *Int. J. Elect. Eng. Educ.*, **12**, pp. 374–384 (1975).

[4] Kane, G. R., 'Personal minicomputer to aid undergraduate instruction in computer system design' *IEEE Trans. on Education*, **E-20**, 1, pp. 10–13 (1977).

[5] Zaky, S. G., Vranisec, Z. G. and Hamacher, V. C., 'On the teaching of computer organisation to engineering undergraduates' *IEEE Trans. on Education*, **E-20**, 1, pp. 27–30 (1977).
[6] Brewer, M. A., *Design Automation of Digital Systems, Vol. 1*, chapter 3, Prentice Hall (1972).
[7] Hayes, G. G., 'Computer-aided design: simulation of digital design logic' *IEEE Trans. on Computers*, **C-18**, 1, pp. 1–10 (1969).
[8] Biancomoni, V., 'Logic simulator programs set pace in computer-aided design', *Electronics*, **50**, 21, pp. 98–101 (1977).
[9] Ashby, D. H. and Johnson, D., 'Simulating digital logic networks', *Electronic Engineering*, **49**, 598, pp. 71–72 (1977).
[10] Walker, A. J., *An Interactive Minicomputer Based Digital System Simulator, Internal Report*, Department of Electrical Engineering, University of the Witwatersrand, Johannesburg (August, 1980).

Part 7

NEW COURSES AND TEACHING METHODS

20

AN UNDERGRADUATE COURSE IN REAL-TIME COMPUTER SYSTEMS

H. S. BRADLOW
Department of Electrical Engineering, University of Cape Town, South Africa

INTRODUCTION
The burgeoning increase in the use of computers suggests that, in many instances, they are becoming as much a part of an engineer's tools-of-trade as an oscilloscope. While, in recent years, engineering students at the University of Cape Town have been increasingly exposed to the use of computers, there are still large gaps in their education in this respect. Students have encountered scientific computation in FORTRAN or BASIC, usually on a batch-processing mainframe, although an increasing familiarity with time-sharing terminal operation or desk-top microcomputers is emerging. Electrical Engineering students have usually also undertaken microcomputer projects where the requirement is to implement a particular control or instrumentation task. The problem is, however, that few (if any) undergraduate students are exposed to the use of a real-time multi-tasking computer system which is the very application they are likely to encounter in practice. This is particularly serious for Electrical and Chemical Engineering students where such computer systems are increasingly becoming an integral part of the systems they will be designing, using and maintaining in industry. Process control and telecommunication switching systems are two obvious application areas where this is the case.

The Departments of Electrical and Chemical Engineering at the University of Cape Town have jointly introduced a course to be given to their final-year undergraduate students on real-time, multi-tasking computer systems. This course, run for the first time in July, 1981, was created both because we believed there was a gap in our education, and because of the encouragement and participation of a local industry, S.A. Nylon Spinners (S.A.N.S.) who are a subsidiary of Imperial Chemical Industries (I.C.I.). The course, set up in consultation with S.A.N.S., had the following aims:
(i) To teach the principles and techniques commonly used in the real-time computer systems in industry.
(ii) To illustrate these techniques by requiring the student to undertake a practical project on a realistic computer system.
(iii) To cater for both Electrical and Chemical Engineering students, as both groups are likely to be involved in process control.

To achieve these aims the course consisted of ten formal lectures supplemented by tutorials and a project undertaken on a PDP 11/23 computer system running under the RSX 11M real-time multi-tasking operating system.

FORMAL LECTURES

The designer and user of a real-time computer system are concerned primarily with three aspects of the computer system that are specifically designed to cater for the requirements of real-time operation (which are described by, for example, Martin[1]):

(i) The operating system;
(ii) The software language used for writing application programs;
(iii) The process interface.

Accordingly, the course work covered these three topics and in each instance based the description on a specific example, as discussed below.

A detailed breakdown of the lectures, giving title, description and relevance of the lecture to the course aims, is provided in Appendix C. The ensuing discussions are an overview of the concepts presented and the motivation for the choice of material included in this course.

Real-time operating systems

This is a highly-specialized and complex topic which could not be covered in depth. However, it cannot be omitted entirely, since it defines the environment in which the user must program and, accordingly, the approach taken was to emphasize the principles and features of a real-time operating system (particularly those that distinguish it from other 'types' of operating system) and not to discuss the details of the interaction with it. The operating system chosen for illustration was RSX 11M[2] since it was the one the students used in their projects.

Real-time languages

Motivation for choosing RTL/2   In order to fulfil the aims of the course, the language chosen for application programming had to have a proven record of industrial and commercial use in real-time applications. Moreover, the time constraints demanded that it be relatively simple, since it had to be learnt by a student in a relatively short period. A further complicating factor in the choice of language is the emergence of the language ADA[3], as a result of the U.S. Department of Defense's (DOD) High Order Language Project. This language, with the mammoth backing of the U.S. DOD, is expected to have a major impact on real-time computing. The problem is, however, that compilers are not expected to be commercially available until the middle of the decade. Thus, whatever language is chosen at the present time, one is faced with the prospect of teaching a language that will be obsolete in the not-too-distant future. It is important, therefore, not to teach, in the interim, a language whose concepts are totally alien to those of ADA. This eliminates languages such as FORTRAN and CORAL 66, since in evaluating existing languages[4] the DOD found these to be 'not appropriate'. (BASIC, which was not evaluated, presumably also falls into this category).

The language RTL/2 was chosen for this course as it fulfils the criteria discussed above. It is used throughout the I.C.I. organization and thus has widespread usage in the process control industry. (It was developed by I.C.I. specifically for this purpose). There is only one version of the language, as it has been frozen since 1973, and thus the confusion arising from variants has been avoided. Furthermore, the language was available and there is local expertise in South Africa.

Teaching RTL/2 In teaching the language in the very short time available, the fact that all students had high-level language experience (using FORTRAN or BASIC) was capitalized on. The motivation for various aspects of the language (such as, for example, the control statements) was therefore considered to be obvious, and so the background discussion was eliminated and the presentation limited to syntax definition. Furthermore, a tight, logically linear sequence of presentation was abandoned, as the use of some feature of the language could be implied without that feature having been formally discussed. For example, the addition of two real numbers could be used as an illustration of a procedure before the use of operators is described, since anyone with high-level language experience would regard this as a perfectly natural operation. The approach, therefore, was to group all subjects that were logically associated under the heading of a single section and present them as a complete topic. As always, a certain amount of compromise was required but this approach was largely achieved.

Since RTL/2 is the first block-structured language the students had encountered, certain topics required greater depth of presentation than others. This is particularly true of the static and dynamic storage of variables and the consequent 'scope' rules for accessing variables. Fixed-point arithmetic, which is an important aspect of RTL/2, also required detailed description, as did features such as 'reference' variables (or pointers) and records.

Although an excellent Training Manual[5] is available for RTL/2, the approach of this course is sufficiently different to make the book unsuitable for use as a student text. Accordingly, a new text[6], which draws heavily on the training manual, was specifically written for this course, the aim being to provide students with a concise explanation of the language for learning purposes as well as a reference manual for use when programming. The scope of the motivating discussion was often limited, as explained above, but the syntax was presented in its entirety, even for those features of the language that the student was unlikely to encounter. The rationale behind this was to enable the student to concentrate initially on the important concepts and, as his confidence and skill grew, to cater for his expansion into greater depth of usage of the language.

Process Interfaces
A Media Process Interface System, manufactured by GEC-Elliot Process Instruments Limited, was chosen to connect the PDP 11/23 to chemical plant located in the Chemical Engineering laboratories. This choice was motivated largely by the expertise and support available locally but was also influenced by the fact that it is a soundly-engineered, industrial system with an established usage.

Consequently, when process interfaces were discussed as part of the course, this Media system was referred to for illustrative purposes. The time available limited the material presented to a brief overview, and so, in order to stress the subject from the industrial users' viewpoint, the topic was dealt with by an engineer from S.A.N.S. who has considerable experience in implementing such systems.

PRACTICAL WORK

The objective of the project is that the student should be able to design and program a control system on the computer to control an item of plant which will involve both batch and continuous control. As the process interface system was not available during 1981, for this year only, the project involved the control of a real-time simulation of a tank as shown in Fig. 1. The odd non-linear shape of the tank was chosen to allow students to apply the control theory they were learning in a concurrent course, if they so wished. The manner in which the tank is controlled is virtually identical to that of the real system, since the information concerning its level and state of valves is stored in a common data base which was updated at 1 sec. intervals by the simulation task, in the same way as a scanning task would update the data base for a Media system.

The project requirement was for each student to write three concurrent, asynchronous, real-time tasks which comprised the control system for the tank.

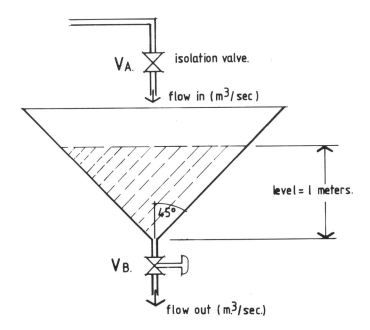

FIG. 1 System to be controlled by students.

These tasks were: (i) to allow the operator to examine values relating to the tank, to change the setpoint, and to switch input flow on and off by opening and closing the isolation valve V_A; (ii) to control the level of the tank using a direct digital control (DDC) algorithm to manipulate the output flow control valve V_B; (iii) to log, on disk, the values of level and output flow with time when a step change in input is applied.

When this project was set, great care was taken to ensure that the importance of the interactions was not obscured by the complexity of the details. This was achieved by providing the student with various learning aids and relieving him of the need to learn unnecessary details. Students were required to use the computer both as a development station and as a target system for performing control. A simplified guide to the timesharing use of RSX 11M and to the editing, compiling and file manipulation utilities was provided to assist with the former. The latter required the student to use certain features of the operating systems. As the details of such interactions were not discussed, routines were provided at compile time, which enabled operations such as securing and releasing the data base or delaying a task for a defined period of real-time to be achieved by means of a simple procedure call. Examples of such routines are given in Appendix A.

A further teaching aid was to provide the student with a program file which contained the guidelines of the programs that he was required to write. This 'shell' program is shown in Appendix B, and had a number of advantages:
(i) It required the student to produce orderly, well-documented programs.
(ii) It reminded the student of the various components that he would require in his program (such as external procedure definitions).
(iii) It supplied features which are essential to the running of a program under RSX 11M, such as the name of the main procedure (RRJOB) and the procedures for opening I/O channels.

DISCUSSION

In order that the effectiveness of this course could be evaluated, the 28 Electrical Engineering students who completed the course this year were asked to complete a questionnaire, rating various aspects. All questions required an answer on a scale of 0 to 10. The average results (indicated by the arrow) and their standard deviation (indicated by cross-hatched bars) are shown in Fig. 2.

From the results of Question 1 there is clearly a demand among our Electrical Engineering students for a course of this nature. Furthermore, the course has, to some extent, achieved its aims in that it has elevated the students' confidence to tackle an industrial process control problem (Question 4). From the students' viewpoint, the use of RTL/2 was a successful choice because they clearly found it superior to the languages they are already familiar with (Question 2), and are moderately confident of their ability to use the language. (Question 3 — this self-assessment of their ability is largely borne out by the examination results). Finally, the importance of the practical work in a course of this nature is borne out by the students' response to it (Question 5).

1. Relative importance of the course in curriculum (rating of present amount of effort devoted to topic)

2. Value of RTL/2 (relative to other languages the students are familiar with)

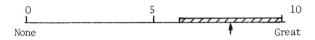

3. Ability to program in RTL/2

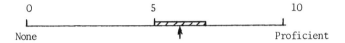

4. Confidence to tackle industrial process control problem

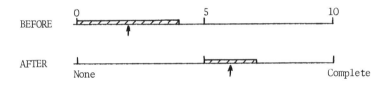

5. Importance of project relative to formal teaching

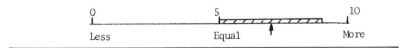

FIG. 2 Student evaluation.

CONCLUSION

The experience gained in 1981 has shown that it is practicable to mount a course of this nature at undergraduate level and achieve a satisfactory student response.

The language, RTL/2, which was used to illustrate the features of real-time programming, was chosen because of its proven industrial value. This course has also proved it to be equally useful as a teaching language as the language was taught with a relatively short amount of formal tuition, and students were nevertheless able to successfully write control systems programs in RTL/2. It also teaches students the desirable features of a structured programming language.

In future years, we intend to continue and expand this course and, when the process interface is delivered, enable students to perform control of running plant.

ACKNOWLEDGEMENTS

The author wishes to thank S. A. Nylon Spinners (Pty) Ltd. for their generous financial assistance in procuring the computer equipment required for this course, for their advice, ideas and support in establishing the course, and for allowing Mr. Stan Cleary (who lectured so ably on Media) to participate in the course. The author also wishes to acknowledge the generous donation of the RTL/2 compiler and support software by H. le Roux and Associates (Pty) Ltd.

REFERENCES

[1] Martin, J., *Design of Real-time Computer Systems*, Prentice-Hall (1967).
[2] *PDP 11 Software Handbook*, Digital Equipment Corporation (1978).
[3] *ADA Programming Language Military Standard, MIL-STD-1815*, U.S. Department of Defense (1980).
[4] Barnes, J. G. P., 'An overview of ADA', *Software Practice and Experience*, **10**, 851–887 (1980).
[5] *RTL/2 Training Manual*, Imperial Chemical Industries Ltd. (1974).
[6] Bradlow, H. S., *RTL/2 User Guide*, Dept. of Elec. Eng., Univ. of Cape Town (1981).

APPENDIX A.   *Routines to interface to operating system*

```
%     EXTERNAL PROCEDURE DEFINITIONS.                                      %
%     ================================                                    %

EXT PROC () RRNUL;                          % NULL PROCEDURE               %
%     THE FOLLOWING ARE RSX 11M EXECUTIVE DIRECTIVES                       %
EXT PROC (REF TASKBUF) RSXGTS;
EXT PROC (INT,INT,INT,PROC()) RSXMRK;       % MARK TIME DIRECTIVE          %
EXT PROC (INT) RSXWTS;                      % WAIT FOR SINGLE EVENT        %
EXT PROC (INT) RSXCLE,RSXSET;               % CLEAR,SET EVENT FLAG         %
EXT PROC (REF FLAGBUF) RSXRDA;              % READ ALL EVENT FLAGS         %

%     LOCAL DATA BRICK DEFINITIONS.                                        %
%     =============================                                       %

DATA OPINT;
    PROC () DUMMY:=RRNUL;
    TASKBUF TTASK:=(0,0,0,0,0,0,0,0,0,0,DUMMY,0,0,0,0);
    FLAGBUF FLAGS:=(0,0,0,0);
ENDDATA;

DATA FACILITIES;                            % DATA BRICK FOR               %
                                            % STORING FACILITIES INFO.     %
    INT FACFLAGS:=0;                        % FACILITY FLAGS - TELL WHICH  %
                                            % FACILITIES TASK HAS SECURED. %
```

```
ENDDATA;
%    LOCAL PROCEDURES.                                                  %
%    =================                                                  %
PROC WAIT(INT TICKS);                    % DELAY TASK FOR TICKS*20 MS   %
    RSXMRK(1,TICKS,1,RRNUL);             % MARK TIME REQUEST            %
    RSXWTS(1);                           % WAIT FOR LOCAL EVENT 1       %
ENDPROC;
PROC SECURE(INT I);                      % SECURE A FACILITY            %
% EVENT FLAG IS RESET IF FACILITY SECURED                               %
%                *****              *******                             %

    INT MASK:= BIN 0000000000000001,
        TEST:=0,EVENT:=33;

    MASK:=MASK SLL (I-1);                % SET MASK APPROPRIATELY       %

    TEST:=FACFLAGS LAND MASK;
    IF TEST   0 THEN                     % FAC ALREADY SECURED BY TASK  %
        ERP(402);                        % ERROR-TASK TRIED TO SECURE   %
                                         % TWICE                        %
        RETURN;
    END;

    EVENT:=32+I;                         % SET GROUP COMMON FLAG FOR FAC %
LABRTR:                                  % THIS SECTION IS TO PREVENT 2 %
                                         % TASKS SIMULTANEOUSLY SECURING %
                                         % THE SAME FACILITY            %

    RSXCLE(EVENT);                       % CLEAR EVENT FLAG             %
    IF RSXDSW   2 THEN                   % EVENT FLAG WAS NOT SET       %
        RSXWTS(EVENT);                   % WAIT FOR IT TO BE SET        %
        GOTO LABRTR;                     % TRY AGAIN TO SECURE          %
    END;
    % FACILITY HAS BEEN SECURED I.E. EVENT FLAG RESET                   %
    FACFLAGS:=FACFLAGS LOR MASK;         % INDICATE TASK HAS SEC FAC    %
    RETURN;
ENDPROC;

PROC RELEASE(INT I);                     % RELEASE FACILITY I           %
% EVENT FLAG IS SET IF FACILITY IS RELEASED                             %
%              ***                ********                              %

    INT MASK:=BIN 0000000000000001,TEST:=0,EVENT:=0;

    MASK:=MASK SLL (I-1);
    TEST:=FACFLAGS LAND MASK;
    IF TEST = 0 THEN                     % A FACILITY NOT SECURED IS    %
                                         % BEING RELEASED               %
        ERP(401);
        RETURN;
    END;
    RSXRDA(FLAGS);                       % READ EVENT FLAGS             %
    IF FLAGS.FLAGCOMMON2 LAND MASK   0 THEN   % FLAG WAS SET            %
        ERP(400);
        RETURN;
    END;
    EVENT:=32+I;
    RSXSET(EVENT);
    MASK:=NOT MASK;
    FACFLAGS:=FACFLAGS LAND MASK;        % CLEAR FACILITY'S FLAG        %
    RETURN;
ENDPROC;
```

APPENDIX B. Outline of RTL/2 programme

```
OPTION (1) CM;
TITLE SHELL;

%                                                                        %
%     DESCRIPTION OF MODULE.                                              %
%     =====================                                               %
%                                                                        %
%                                                                        %
%                                                                        %
%                                                                        %
%                                                                        %
%                                                                        %

%     LET DEFINITIONS.                                                    %
%     ===============                                                     %

LET LF = OCT 012;              % LINE FEED CHARACTER                      %
LET CR = OCT 015;              % CARIAGE RETURN CHARACTER                 %
LET VT = OCT 013;              % VERTICAL TAB CHARACTER                   %
LET SP = OCT 040;              % SPACE CHARACTER                          %

%     MODE DEFINITIONS.                                                   %
%     ================                                                    %


%     EXTERNAL PROCEDURE DEFINITIONS.                                     %
%     ==============================                                      %

%    THESE PROCEDURES ARE USED TO OPEN AND CLOSE I/O CHANNELS             %
EXT PROC(INT,REF ARRAY BYTE) RSXOPI,RSXOPO;
EXT PROC(INT) RSXCL;
EXT PROC(INT)INT RSXSWI,RSXSWO;
EXT PROC() RSXINT;
%    THE FOLLOWING ARE STANDARD STREAM I/O PROCEDURES - REFER             %
%    TO THE STREAM I/O MANUAL                                             %
EXT PROC(REF ARRAY BYTE) TWRT;
EXT PROC (REF ARRAY BYTE,REF ARRAY BYTE)INT TREAD;
EXT PROC () INT IREAD;
EXT PROC (INT) IWRT;

%     EXTERNAL DATA BRICK DEFINITIONS.                                    %
%     ===============================                                     %

%     LOCAL DATA BRICK DEFINITIONS.                                       %
%     ============================                                        %

DATA LOCAL;
    REF ARRAY BYTE TERMB:=" CR,LF ";    % TERMINATING CHARS. FOR INPUT    %
ENDDATA;

%     MAIN PROCEDURE = RRJOB.                                             %
%     ======================                                              %

ENT PROC RRJOB();

    RSXINT();                        % THIS PROCEDURE IS                  %
                                     % TO INITIALIZE THE STREAM I/O       %
                                     % INTERFACE                          %
    RSXOPO(1,"TI:");                 % OPEN TI: AS AN OUTPUT STREAM       %
    RSXOPI(2,"TI:");                 % OPEN TI: AS AN INPUT STREAM        %
%   RSXOPO(3,"LOGFIL.OUT");          %% OPEN FILE FOR LOGGING OUTPUT      %
    RSXSWI(2);                       % SET INPUT STREAM TO TI: KEYS       %
    RSXSWO(1);                       % SET OUTPUT STREAM TO TI: VDU       %
```

```
ENDPROC;
%    OTHER ENT PROCEDURES.                                                    %
%    =====================                                                    %

%    LOCAL PROCEDURES.                                                        %
%    =================                                                        %
```

APPENDIX C. Detailed breakdown of lecture material

Lecture	Title	Description	Relevance
(1)	Introduction to real-time computing	Introduces the concepts of real-time computing by means of an example of a computer controlling a process, and discusses the requirements of and constraints on such a computer system. The concept of standard software categories (such as operating systems) is also introduced.	The lecture defines the problems that the techniques taught in the rest of the course are aimed at solving.
(2)	Features of real-time operating system: Part I	Describes the executive control features and how executing tasks share the system resources.	These concepts explain the software performance of the computer and how it affects control system design.
(3)	Features of a real-time operating system: Part II	Describes the features of the operating system available to the programmer such as system directives, I/O drivers and software development utilities.	These are the features of the operating system which the user requires in order to run his/her program in its environment.
(4)	Introduction to RTL/2	Discusses the features that one desires from a real-time language and the general syntactical structure of RTL/2.	Serves as a basis for the description of the use and suitability of RTL/2 for real-time work. The ensuing 5 lectures describe the syntax, structure and real-time features of RTL/2.
(5)	RTL/2 Data types and declarations	Describes the data types available in RTL/2 and the syntax for declaring them.	The integrity obtained by strict definition of data types is illustrated.
(6)	RTL/2 Procedures and Function Calls	Describes the manner in which RTL/2 procedures and functions pass parameters and use the 'stack' for local (dynamic) data storage.	The concepts of re-entrant code, shared by a number of tasks each with its own local data stack, are illustrated.

(7)	RTL/2 Data bricks and data initialization	Describes how RTL/2 procedures may share common 'static' data bricks which may be initialized at compile time.	Program clarity and integrity are improved by the ability to separate common data into bricks.
(8)	RTL/2 Operators and expressions	Describes how various arithmetic and logical operations are implemented in RTL/2.	The manner in which machine independence of source code is achieved through strict definition of operation on data types is illustrated in this lecture.
(9)	RTL/2 control structures and inter-module communication	The syntax of the control structures available in RTL/2 is described. Also the capability of writing RTL/2 programs in the form of easily separable modules is illustrated.	The usefulness of being able to write modular programs is stressed
(10)	GEC's MEDIA Process Interface System	A brief description of the structure and use of this particular process interface systems is given.	The control system programmer interacts with the outside world by means of this process interface and hence an understanding of it is required.

21

AN ADVANCED ELECTRONICS TEACHING LABORATORY

R. M. F. GOODMAN, G. E. TAYLOR and A. F. T. WINFIELD
Department of Electronic Engineering, University of Hull, England

1 INTRODUCTION

This paper describes a new third-year teaching laboratory, which has been implemented during the current academic year, as an integral part of the four-year honours degree course in Electronic Engineering at the University of Hull.

Formal teaching and laboratory work in the department are continually reviewed, as befits the rapidly changing nature of the subject. As part of this review process, the third year laboratory was closely examined in the summer of 1981, and it was felt necessary to make a number of major changes to reflect both the evolving nature of the technology, and changes in the philosophy of engineering education[1].

This paper proceeds as follows: the next section examines in more detail the reasons for change, and this is followed in section three by a brief examination of the actual laboratory and its experiments. Next, we examine the mode of assessment adopted for the laboratory, and, finally, conclude by reflecting upon the first year of operation.

2 REASONS FOR CHANGE

Traditionally, electronics teaching laboratories have been based upon formal experiments designed to demonstrate specific devices, or techniques, to the student. Clearly, such experiments remain a necessity in the early years of the course, in order to provide the fundamental understanding of the subject which is needed before design skills can be acquired.

Practical work in the final year takes the form of an extensive, open-ended project in which students, with minimum supervision, are encouraged to use appropriate technology to solve original design or research problems. In recent years the techniques used in final-year projects have escalated in complexity so that, for example, original solutions involving purpose-designed microprocessor hardware have become commonplace.

In this context, the third-year laboratory was seen as a transition from the closely-supervised traditional experiments of the second year, to original design work in the final year. To be more specific, the third-year laboratory must incorporate two important features:
(i) experiments involving the new microelectronics technology and
(ii) open-ended experiments requiring greater initiative on the part of the student.

3 THE NEW LABORATORY

3.1 Organisation

At the outset, it was felt that the six hours traditionally allotted to each experiment in the first three years' teaching laboratories would be insufficient for the 'project'-like experiments envisaged. This, combined with the limited physical space available, lead us to propose that students spend alternate weeks in the laboratory, working for up to twelve hours on single experiments. With a total of twenty working weeks in the Autumn and Spring terms, ten experiments were then needed. In addition, single student work stations were to be the norm, students pairing up only where unusually expensive equipment was involved.

In keeping with the project nature of the new laboratory, it was felt that intensive supervision would be unnecessary, indeed, possibly harmful. A reasonable amount of supervision should, nevertheless, be available if required, at the start of each week. Accordingly, the fifteen-work-station laboratory is manned by three staff, and one postgraduate demonstrator for the first afternoon session (of three hours), and then, for the remaining three afternoon sessions, just one member of staff, and one demonstrator. A laboratory technician is also present during the whole twelve-hour period, to deal with any problems with the equipment.

3.2 The experiments

An important consideration, when devising actual experiments, was that they should be multi-layered, so that there is sufficient material for the weak student to become involved and learn at his own level. At the same time, additional ideas are available for further exploration by the more able students.

Appendix 1 contains extracts of the actual laboratory handout sheets for those experiments which, in the authors opinion, best represent the philosophy of the new laboratory.

The following is a breakdown of the subject areas covered in the laboratory, and details of the experiments within each area.

3.2.1 Analogue computation
Despite the advent of powerful digital computers, analogue techniques still provide the quickest solution to certain types of problem. Indeed, there has been a recent resurgence of interest in hybrid computational techniques for the control of complex structures such as robots. This experiment introduces students to the principles and practice of analogue computation via a series of graded exercises leading up to the design and implementation of a patch for a problem requiring simple iterative and hybrid techniques. Able students are encouraged to devise and patch problems allied to their own particular fields of interest rather than being rigidly guided by the laboratory sheet.

FIG. 1   *The teaching laboratory.*

3.2.2 *Communications*   The degree course, in this department, has a significant communications component throughout. Accordingly, we felt justified in including two experiments to investigate advanced communications techniques, one in analogue communications and another in digital communications, both to use state-of-the-art devices and techniques.

The analogue experiment is based upon a full specification SSB transmitter module, available commercially, and intended primarily for amateur application. The experimental transmitter modules have been assembled such that all of the internal signals are accessible for examination. Integral with the experiment is a sophisticated spectrum analyser and the primary objectives of the experiment are twofold; to evaluate the performance of the transmitter module — and verify (or otherwise) the manufacturer's specification — and, at the same time, become familiar with all of the techniques of spectral analysis.

The digital communications experiment is designed to demonstrate the techniques and problems of Pulse Code Modulation (PCM). The experimental hardware consists of four functional modules; an analogue to digital converter, a digital to analogue converter, and two universal asynchronous receiver transmitter ICs (UARTs), set up in parallel to serial, and serial to parallel configurations. The modules may be operated singly, or in combination, the student having to design the interconnection, timing and filtering in order to construct a model serial PCM system.

3.2.3 Design and breadboard It was felt that students in the third year should rapidly acquire the basic skills of design and breadboarding and the ability to work from data sheets and specifications. In addition, we wanted to introduce them to the discipline of getting components from our stores system and negotiating with storekeepers. To this end, two design experiments were evolved, one predominantly digital, the other analogue.

The digital experiment consists of a design specification for an electronic combination lock safe. The objectives are to design and breadboard the system to the working prototype stage. Students are given data sheets on a keyless-lock IC and a hexadecimal keyboard as a starting 'hint'. From then on they can draw any components they need to build up the system. The Radio Spares catalogue is also available, and students are expected to consult data books and data sheets in the departmental library as needed. In this way, the organisational and practical problems of getting a design from paper to breadboard are rapidly brought home.

The analogue experiment is centred around a number of analogue integrated circuits: the 3900 Norton amplifier, 555 timer, and phase lock loop, frequency to voltage converters etc. Students are given the data sheets on these devices and asked to investigate some of the suggested applications shown. An objective of the experiment is to design a 'system' using analog ICs, and an example is suggested which is a gas sensor alarm — telemetry system. Students are, however, encouraged to think of new applications and design and build their own system.

3.2.4 Microprocessor interfacing This experiment aims to teach the student the basic techniques of interfacing a microcomputer to the real world, something that is often neglected in 'traditional' microprocessor courses. Each workstation has an Apple II microcomputer, with disk drive, and a variety of peripherals are available. The advantage of using a sophisticated microcomputer as the controller, as opposed to a lower-level microprocessor development system, is that the student can work in both a high-level language and a machine code environment, and quickly develop powerful control algorithms. The peripherals that are available include a Smart-Arms robot, a general purpose input-output module with push button switches, indicating LEDs, and analogue potentiometers, and a set of commercial interfaces manufactured by Feedback. These include a simulated washing machine, a simulated traffic lights/pedestrian crossing system, a heating control system, and a stepper motor module. Students are presented with documentation on the Apple and peripherals and are expected to choose what they wish to do. The excellent documentation supplied with the Apple II enables weaker students to quickly brush up their BASIC with a 'tutorial', whilst stronger students are expected to design their own interfaces and peripherals. To date, two such interfaces have been designed by students on their own initiative, a light pen, and a graph plotter interface.

3.2.5 *Microwaves*   The unusual nature of microwave hardware and devices means that students rarely have an opportunity to experiment first-hand. This single experiment is designed primarily to provide this opportunity, and is based upon the Sivers microwave components — designed particularly for educational use. The hardware consists of a Klystron as a microwave source, and a selection of microwave devices and waveguides which may be bolted together to form a complete working system. The student is asked simply to become familiar with the operation and use of each device, and finally verify a number of fundamental relationships, for example, the relationship between free-space and waveguide wavelength.

3.2.6 *Software*   The laboratory includes two experiments based on the production of computer programs. This reflects the authors' view of the increasing importance of software techniques in electronic engineering. The first experiment is documented in Appendix 1.1 and aims to encourage students to produce not just a working program, but a fully-documented piece of software of an industrially acceptable standard.

Computer simulation is a technique used increasingly by the electronic engineer as a method of design verification, before any effort is committed to actual hardware. The second software-oriented experiment is designed to demonstrate the use of computer simulation. The student is asked to model mathematically a circuit, or system, either real or imaginary, and produce from this model a working simulation capable of verifying, or otherwise, the original system specification. The key requirements of this exercise, are that the model is an accurate representation of the specified system, and that the final program truly simulates the model.

3.2.7 *Systems analysis and control*   This experiment seeks to introduce the student to sophisticated network analysis equipment and also to provide a practical basis for work covered in the third-year Automatic Control course. It is in two parts each of which is expected to occupy about six hours. The first part allows the student to gain familiarity with programmable network analysis equipment (the system comprises a Hewlett-Packard desk-top computer controlling a frequency synthesiser and gain-phase meter) and to use frequency response techniques to identify an unknown linear system. The second part requires the modelling and control of an analogue system (a d.c. servo) and a comparison between theoretical and measured behaviour. During the course of the experiment, students are encouraged to make full use of a set of interactive control system design programs available on the departmental PDP11/34 computer and thus gain further insight into the role of computer-aided design in modern engineering practice.

4 ASSESSMENT

Assessment in the new laboratory takes the form of formal laboratory reports submitted for each experiment. In addition, students are required to keep day books as personal records of work done. Formal reports are written in the alternate 'week-off' between laboratories.

Whilst this is not itself a departure from standard practice, the actual format of formal reports required has been radically altered (Appendix 2). In particular, it is felt that the formal report should not follow a standard pattern but should more closely reflect the actual nature and purpose of the work carried out. For example, in the case of the 'Production of technical software' (Appendix 1.1), the formal report comprises a fully-commented listing and users' and programmers' guides.

5 CONCLUSION

The new laboratory is now approaching the end of its first year of implementation, and appears to have been a major success. We have noted a greater response from students, in developing their own ideas within the framework of the experiments, from the whole range of abilities within the class. In particular the 'layering' of experiments seems to work well, in that the weakest students do not experience difficulties, whilst the more able students do not lose interest, as is often the case with traditional tightly-controlled experiments.

REFERENCES

[1]   The Finniston Report: *Engineering Our Future, Report of the Committee of Inquiry into the Engineering Profession, Command Paper No. 7794*, Her Majesty's Stationary Office, London (1980).

APPENDIX 1.1

Experiment 1   Production of technical software

Objectives   The aim of this experiment is to write and document to an industrially acceptable standard a piece of software to solve an engineering problem.

The Problem   A large number of different problems have been selected and students will be allocated one of these at random at the beginning of the experiment. However, any student wishing to tackle a problem of his own devising is encouraged to do so, provided he first discusses the project with Dr. G. E. Taylor.

The Programme   Choice of level and type of language and of machine is left to the student, but some justification of this choice should be made in the Programmers' Guide (see section on documentation). *Among* the factors to be considered are
(i)   portability of high level languages such as FORTRAN, PASCAL and ALGOL
(ii)   speed of low level implementations
(iii) general, multi-user accessibility of main frames
(iv)   usefulness of programs to run on personal computers since these are now widespread
(v)   desirability of graphical output
(vi)   machines available! — any university or departmental computer normally available to undergraduates may be used, alternatively a student may use his own machine, but see note on assessment.

Programs for programmable calculators are acceptable provided that the writer demonstrates the usefulness of solving the given problem on such a machine.

Documentation The source file itself should be *fully* commented unless this is impossible as with a program for a programmable calculator in which case a list of codes plus comments should be supplied. In addition two pieces of documentation are required.
(1) A Users' Guide written in such a way as to enable an engineer who is not conversant with computers to understand the task performed by the program, to enter his own data successfully, run the program and interpret the output.
(2) A Programmers' Guide detailing *briefly* and *clearly* the algorithm, data structure and so on to enable another programmer to understand and amend the software.

Assessment A standard formal report is not required. The work will be assessed from an examination of the Users' and Programmers' Guides and the commented source together with a demonstration run of the program.

APPENDIX 1.2

Experiment 5 — A P.C.M. link

1 *Objectives and Overview* The key elements of any communication system employing Pulse Code Modulation, are the analogue to digital conversion (A–D), and digital to analogue conversion (D–A), the modulation and demodulation processes respectively.

Practical A–D and D–A convertors usually have parallel digital interfaces, and so the additional processes of parallel to serial, and serial to parallel conversion are incorporated to give a serial pulse coded bit stream.

This experiment models PCM modulation and demodulation using off the shelf MSI components; the ZN425E 8 bit A–D/D–A converter, and the 6402 Universal Asynchronous Receiver Transmitter (UART). Two of each of these devices are supplied, so that the individual devices may be evaluated in each of their dual modes — and ultimately all four may be interconnected, to model a complete PCM link.

The objectives of the experiment are:
(a) To gain familiarity with these devices.
(b) To evaluate the devices over the complete range of operational parameters.
(c) To appreciate the problems, and concepts of Pulse Code Modulation.
(d) To model a PCM system.

2 *Experimental* A suggested programme of experimentation is as follows:
(a) *A–D, D–A Converters*
   Test the converters separately using DC input voltages, and binary switches. Use the Farnell pulse generator to generate single TTL level pulses, to 'single step' the A–D converter.
   Calibrate and test for linearity.
   Link the digital output of the A–D to the input of the D–A, and compare analogue input of the A–D, with analogue output of the D–A.
   Clock the coupled A–D/D–A over a range of speeds, and analogue waveforms. See the enclosed diagram showing recommended signal generators.
   Define the useful ranges of clock, conversion and analogue frequencies.
   Observe the effects of 'stuck at' bits in the digital interface, and the effects of exceeding the analogue frequency limits.
   Design and build an output filter for the D–A converter.
(b) *UARTs*
   Single step the UARTs. Observe the serial output for different parity, length and stop bit

options. Generate a serial bit stream manually, and clock this into the serial to parallel UART.
  Transfer an 8 bit word across the interface by manual clocking.
  Observe the effect of mismatching the parity, length and stop bit options.
  Clock the UART pair at high speed, and observe the bit stream using an oscilloscope and logic analyser combination.

(c) *The PCM link*
  Connect, using the ribbon cables, the units to form a serial PCM link. Run the link over a range of speeds, and analogue input waveforms (including speech), observing the digital data using the logic analyser.

Notes: Please ask a demonstrator to check your power connections to A–D/D–A and UART boxes before you switch on. *Ensure* that signal generator outputs used for clocking and control are TTL levels (0–5V).

3 *Assessment* The formal report should be an assessment of suitability, for application in an actual PCM communication system, of the ZN425E and 6402 devices. Include a design for a control circuit, to provide all of the timing and control signals, for the system.
  Inclusions: Circuit diagrams and data sheets.

APPENDIX 1.3

Experiment 8 Apple interfacing

Introduction This experiment is all about controlling peripherals with the Apple II microcomputer. You are provided with a number of interfaces to the Apple, and a number of peripherals. These include:
  (1) Buffered version of the Apple's Games/I/O Connector.
  (2) Simulated washing machine.
  (3) Simulated traffic lights.
  (4) Heat control system.
  (5) Stepper motor.
  (6) Smartarm robot.
You will have to make full use of the Apple and peripherals documentation provided.

What you are going to do You are going to control the various peripherals by writing control programs in BASIC.

How to go about it
  (1) You must be able to program in BASIC. If not, learn quickly by reading the APPLESOFT TUTORIAL.
  (2) Learn how you can make I/O lines go 'high' or 'low' by use of the POKE command. Also how you can sense the state of an input line by use of the PEEK command. Use the Buffered games I/O connector, turn output LEDs on and off, sense the state of the input lines and the angle of the pots. Use the Applesoft Manual and the Apple Reference Manual.
  (3) Choose one of the FEEDBACK peripherals, read the manual and write a BASIC program to control it.
  (4) Control the Smartarms Robot. Get it to do a simple pick and place task. Then more complicated tasks if you wish. Remember you must share the robot.
  (5) *Either* (a) Try some of the other FEEDBACK peripherals, *Or* (b) Design and breadboard your own peripheral, interface it to the Apple via the buffered games I/O connector, and control it from a BASIC program.

Hints:
(1) Keep your programs simple and well documented.
(2) You can save a program by typing SAVE filename. Load a program by typing LOAD filename. Don't get bogged down in the Apple's DOS.

(3) Double check power connections.
(4) Don't blow up the Apple.
(5) You can take the documentation away during your week 'on', and the subsequent report week. If you lose it we will charge you.
(6) Don't be afraid to do your own thing in this experiment — but discuss it with a demonstrator first.

Report Requirements

(1) Your report should tell me what you did and what the results were.
(2) Don't tell me about the Apple, peripherals, robot. I know more than you about these and don't want to read chunks of regurgitated manual.
(3) Include listings of your programs, and circuit diagrams of your *own* peripherals.
(4) Keep it short.

APPENDIX 2

Guide to writing formal laboratory reports

1 *Overview*   The preferred style and format of laboratory reports in the third year lab is that of a technical report. This will be a significant departure from what you have become used to in previous years, and will approach the type of report required in the real engineering world. Imagine the laboratory experiment as an evaluation of a new equipment, or technique or concept — and you are then to report the results of your evaluation (to a project manager, for example). In a real situation you would be testing out the equipment for possible application to an engineering problem — and on the basis of your technical report the project manager would make the decision of whether to incorporate the equipment or not. The technical report should then be:
    (i) Brief. Your time, and your project managers' time is valuable.
   (ii) Accurate. Your project manager is relying on *your* report to make his decision (which could be expensive) — so get your facts right.
  (iii) Descriptive rather than mathematical. If you have to quote mathematical results include a verbal description explaining what the mathematics *really means* in engineering terms. Your project manager may not be an expert in the field, that's why he employed you to do the evaluation! If you must include proofs put them in an appendix.
  (iv) Informative. Don't tell your project manager what he knows already — and don't include a blow by blow account of how you undertook your evaluation, he's not interested!
   (v) Conclusive. Don't leave your report open ended — otherwise you'll be asked to write another.

2 *Format*   The main body of your report should not be more than 5 sides of A4, excluding graphs, diagrams, appendices and references. Reports significantly longer than this will be penalised.

The actual format will of course depend on the nature of the experiment, but will probably consist of:
  (i) *An Abstract*   This, *very* briefly states what the report is about — the object of the work, and major conclusions of the report. (Most people decide whether to read a paper or not, according to the abstract).
 (ii) *Introduction/overview*   It is difficult to generalise upon this/these sections, because this part will depend very much upon the actual work, but here you will 'fill in the background' needed to appreciate section (iii). (i.e. a theoretical summary).
(iii) *Results summary and discussion*   This is the most important part of the document — and contains the actual findings of the work — with reference to results (experimental or theoretical). The results are examined from an engineering viewpoint.

(iv) *Conclusions* Using the previous analogy the conclusions would make *actual* recommendations. (i.e. is the technique useful/applicable/relevant/reliable etc?).
(v) *References*
(vi) *Appendices* (where applicable). Put detailed results, and working in here, if you feel it must be included.

3 *Presentation* Almost any presentation is acceptable (except loose sheets), provided it is *legible* without the assistance of a handwriting analyst!

4 *Submission commitment* A Report is required for *every* experiment, and will be collected by the laboratory technician on the first day of the subsequent experiment. (Every two weeks).

Reports one day late will probably suffer a penalty, any later and you risk zero marks.

5 *Special requirements* Most of the experiment sheets will contain a note with specific requirements for the formal report.

22

MICROPROCESSOR ENGINEERING AND DIGITAL ELECTRONICS: AN EXAMPLE OF AN M.Sc. COURSE AT UMIST

M. G. HARTLEY
Department of Electrical Engineering and Electronics, University of Manchester Institute of Science and Technology, England

BACKGROUND
In the autumn of 1978 the author of this present paper prepared a contribution to the Pacific Region Conference on Electrical Engineering Education, held in Adelaide in December 1978 which was subsequently published in IJEEE with the title *An Example of an M.Sc. Course — Digital Electronics at UMIST*[1]. The paper dealt with various aspects of an M.Sc. Course in digital studies and its evolution over the years since its modest beginnings in the mid 1960s. Further change, with a new title incorporating the word *microprocessor*, was predicted for the course in the near future.

Since the preparation of this paper, substantial change has taken place in respect of the course. A fresh title *Microprocessor Engineering and Digital Electronics* has reflected new emphasis in subject material. Staff from the Department of Computation have joined colleagues from the Department of Electrical Engineering and Electronics in offering lectures and laboratory activity in the software area, to the extent that the collaborative effort has resulted in a joint course.

In view of all these developments, an updated article seems appropriate. The author of this paper writes in his capacity as Course Organiser over the past five years, but must stress that opinions offered are his own.

OVERALL SCHEME FOR THE COURSE
The course began as the *Applied Electronics M.Sc. Course* in 1964, with only a few students. The title was modified to *Digital Electronics* as early as 1969 to reflect the growing importance of computer science and digital technology. A further change ot title *Microprocessor Engineering and Digital Electronics*, (M.E.D.E.), came in 1979. This change coincided with the enhanced collaboration with the Department of Computation as a result of which the course is now offered jointly by the two departments. This collaboration was facilitated by an appointment in the Department of Computation of a professor with microprocessor engineering interests in 1978, and two further professorial appointments in 1980, one a chair of software engineering in the Department of Computation and the other a chair in applied microelectronics in the Department of Electrical Engineering and Electronics. Associated with these professorial appoint-

ments have been further posts in both departments at lecturer level. In addition, a limited number of support staff have been appointed.

Over the last decade, numbers on the course have been between 20 and 35. Until the recent past, some students from Brunel University and Bradford University have attended the lecture content of the Course for the first two terms, Autumn and Spring, each of 10 weeks, before returning to their parent university for the project, which, in most cases, is completed by the end of the twelve-month period. Statistics as to numbers are given in Table 1, from which it will be seen that numbers in recent years have been of the order of 30 per year. The table also provides details of success rates for the course. Over the years the percentage of passes at the examinations, held in May and relating to the work of the taught part of the course, has been at a consistently high level of about 90%. The percentage of students successfully completing their dissertation has been rather less, with about 80% of the original intake of students graduating. The subject of the dissertation phase of the work is important and will be referred to again later in this paper.

OBJECTIVES OF THE M.E.D.E. COURSE

Various branches of engineering are undergoing change at the present time. Development is particularly rapid in computer technology and especially in areas related to microprocessors and their applications.

Undergraduate courses in electrical engineering, electronic engineering, physics and computer science must provide a wide spectrum of material and, of their very nature, are slow to respond to rapid syllabus change. However, there is a considerable need in industry for young graduates who are sufficiently knowledgeable in the area of digital techniques and microprocessor technology to be able to make an immediate contribution.

The aim of the M.Sc. course is to build on a conventional undergraduate course in electrical engineering, electronic engineering, computer science or physics in order to train young engineers to make such a contribution. Ideally, such an engineer must be capable of guiding a project from initial customer specification, through design and prototype manufacture to evaluation and test. Throughout the development the various trade-offs between software and hardware aspects of the project must be kept in mind, and cost considerations appreciated. A training at M.Sc. level to achieve fully such an objective is difficult of realisation. In many cases full appreciation of all the aspects of the design process will only become apparent after graduation. However, it is felt important in the course to present the basic techniques, to indicate their interrelations and to foster a systems approach, not least through the dissertation project.

The approach follows the traditional pattern of two terms of lectures, tutorials, seminars and laboratory work with four examinations at the end of the second term. A dissertation project forms an important part of the course. This begins as a feasibility study in term two. During this period, the overall project is assessed, preliminary designs are sketched out and cost is considered.

TABLE 1   M.Sc. course statistics. Applied Electronics, 1964–69; Digital Electronics, 1969–79; Mircoprocessor Engineering and Digital Electronics, 1979 to present.

YEAR	NUMBER OF STUDENTS	COMMENTS ON MAY EXAMINATIONS	TOTAL OF GRADUATES
1964-1965	5	3 Fails	
1965-1966	10	1 Fail	5
1966-1967	10		13
1967-1978	10	3 Fails	23
1968-1969	12	2 Fails * Resat 1970	30
1969-1970	14*	4 Diplomas (1st Year for Dip.)	44
1970-1971	13	1 Diploma	51
1971-1972	18	1 Diploma	69
1972-1973	20 + 1		85
1973-1974	11 + 3	2 Diplomas	103
1974-1975	20	1 Fail	116
1975-1976	17		128
1976-1977	24		149
1977-1978	15	1 Fail	170
1978-1979	14		183
1979-1980	19	First year as MEDE	200
1980-1981	32	3 Diploma 2 Fail 1 Special Student	
1981-1982	29*		

Notes: (i) The total number of graduates comprises those students who, after passing their May Examinations subsequently present a dissertation thesis acceptable to the examiners. (ii) Details as to students from Bradford and Brunel Universities have been omitted from this table. (iii) *1981–82 includes one part-time Module I student.

Work on the project begins full-time after the examinations, held early in May and normally continues until October. The final date for submission is 15th October. The dissertation thesis and its preparation provide a valuable foretaste of what is to be expected in the industrial environment upon graduation. This is especially the case since considerable importance is attached in the dissertation work to a careful assessment of objectives, alternative strategies for problem solution and careful construction of hardware, where appropriate. Robust software, properly documented, is equally important, as is also a lucid written presentation of the project in the dissertation thesis.

STUDENT INTAKE

Normally students are recruited with a background in electrical engineering, electronics, physics or computer science. Occasionally students are admitted with qualifications in other technical fields or in mathematics. In these cases the interview, which is provided for all potential students who are already in the U.K., provides an opportunity to assess the aptitude of the student to profit from the course.

At the present time the numbers on the course are divided approximately equally between home students and overseas students. One significant feature of the recruitment of home students is that many are mature students up to the age of 40 or even in their 50's who are retraining as a result of losing their jobs through redundancy or early retirement. For these students the Manpower Services Commission through its TOP's scheme (Training Opportunities Programme) sometimes provides financial support as well as fees. For the younger candidate there is the possibility of an award from the Science and Engineering Research Council (SERC). These SERC awards fall into two classes. One is intended for recent graduates and operates under a quota system; hence the term Quota Award. The other, not subject to the same restrictions, is the Instant Award designed for those with several years' industrial experience. One continuing source of U.K. students has been staff of technical colleges and polytechnics who can sometimes secure secondment for a year as 'mid-career training'. Such candidates can offer much to the course by their mature attitudes to their studies.

The overseas students continue to come from certain traditional areas of the world; Latin America, Portugal, Greece, the Middle East, Indonesia, Singapore and Hong Kong, with occasional students from Africa, New Zealand and European Countries. Many are funded by governmental grants or are seconded by the universities at which they are young staff members. Others are funded privately, frequently by members of their extended family. The considerable increase in fees payable by overseas students imposed by the U.K. government over recent years is a very substantial burden for these students. As a result of these increases, U.K. fees are now among the highest in the world, equalling those of the prestigeous private universities in the U.S.A.* The fees issue is an

---

*There are, however, two chinks of light in the overseas fees position. The first is that students from the European Economic Community (EEC) pay only the U.K. fees under reciprocal funding arrangements. As a result, course organisers as well as potential students await the entry of new members to the E.E.C. with very keen interest, since the result is a reduction of student fees from about £4000 per year to the much lower figure of about £1,200 per year. The second possibility of a reduction of fees is via the special support scheme for particularly able overseas students, by which a remission of fees to the U.K. level is possible. This scheme is administered by the Committee of Vice Chancellors and Principals (CVCP).

emotive one with U.K. university educators and has been discussed in a recent Journal Editorial[2] and there is no sign of any shift in governmental attitudes. A further factor to discourage overseas students is the high cost of living in the U.K. Certainly there is evidence that, for students, even the U.S.A. is a cheaper place to live than the U.K. In view of these discouraging circumstances it is very agreeable to report that the MEDE Course receives considerable support from overseas candidates of good quality.

In the earlier article[1] reference was made to students attending the lecture section of the Course from the Universities of Brunel and Bradford. For some years now Brunel University has run its own M.Sc. Course and accordingly has withdrawn from the earlier collaboration. Bradford University began its own M.Sc. Course very recently and with the current session, 1981–82, the final Bradford student attends the Manchester Course.

MODULAR ARRANGEMENTS

With the session beginning in October 1981, a very serious attempt was made throughout UMIST to offer all M.Sc. courses, and there are many, in modular form. The intention was two-fold. One was to permit students to attend the taught content of an existing course on a part-time basis over a period of up to four years. The other was to allow students in appropriate subject areas to build up modules from more than a single existing course. In this way, the intention was to permit students to prepare themselves in a field of special interest or relevance to their own career.

Accordingly, most M.Sc. courses are now offered on the basis of four modules in which Monday, Tuesday, Thursday and Friday are treated as separate units each comprising integrated courses of study consisting of lectures, tutorials and laboratory and extending over 10 weeks in the Autumn term (Michaelmas term) and a further 10 weeks in the Spring term (Lent term). Each module is examined separately and the project can begin when perhaps two modules have been completed. Nevertheless, the typical time to complete the course would be five years. This is a formidable undertaking.

As reported earlier, several M.Sc. courses in the Department of Electrical Engineering and Electronics at UMIST have relied, over the years, on the M.E.D.E. Course to provide substantial contributions to their lectures and laboratory work. The advent of the modular structure has provided some opportunity to formalize existing arrangements. Accordingly, the M.Sc. Courses in Communication Engineering and in Power Electronics take the Monday module with the title 'Microcomputer Engineering', while several individual lecture series are taken by the M.Sc. Course in Instrument Design and the M.Sc. Course in Integrated Circuit System Design. This latter course has been set up only recently (October 1980) with the aid of special funding from the SERC and represents a collaborative effort between the Microprocessor Engineering Unit, the Digital Processes Group and the Solid State Electronics Group of the Department of Electrical Engineering and Electronics, UMIST, the Department of Computation, UMIST, and two local manufacturing companies, Ferranti Ltd and International Computers Ltd[3].

So far, the modular arrangements have had a disappointing response from potential students. While it is difficult to pinpoint precise reasons, it is clear that the current depressed state of manufacturing industry throughout Britain, coupled with the withdrawal of industrial enterprises from the North West region are substantial contributing factors to the lack of interest in modular courses. Indeed, applications for the M.E.D.E. course in modular form have come, in the main, from those employed as lecturing staff in higher education in technical colleges who are sometimes able to reschedule their lecture programme to leave one day free each week for modular studies.

The provision of courses on a modular basis provides substantial problems for course organisers. One fundamental difficulty is the attempt to split a course which has been designed in an integrated manner into four sections, each of which may be regarded as intellectually satisfactory and which may be taught without reference to the other three. Compromise is inevitable, and, as an absolute minimum provision, a 'preferred order' in which to attempt the modules must be specified to the intending students. Modular arrangements have a profound influence on timetable constraints and restrict any changes to within the relevant day. Flexibility in the use of scarce laboratory resources which may be restricted by a modular approach is a further hazard which must be remembered when contemplating modular schemes. Certainly such schemes should not be undertaken lightly.

COURSE CONTENT

The earlier article gave details of course content for the Session 1978–79. This present paper discusses details for the Session 1981–82. The differences after only three years will be readily apparent when the timetables shown in Figs. 4 and 5 of the earlier article are compared with the timetables shown in Figs. 1 and 2 of the present paper.

The modular nature of the course is further illustrated in Table 2. In each case the module comprises lectures with supporting laboratory or programming work. Wednesday morning is kept for occasional seminars and tutorials while Wednesday afternoon is the time traditionally reserved in the UK university system for sporting activities though it is doubtful if the students on the course use the time for anything other than private study. The Monday module is regarded as the 'foundation module' and students are required to complete this module first. Thereafter it is likely that they would be recommended to carry out the Thursday module on Software Engineering.

Monday module: Microcomputer Engineering
Since students come to the course from diverse backgrounds, the first five weeks of the course include some material of a revision nature which is used to bring all students up to a common level as soon as possible. Accordingly, a two-hour slot on the first five Monday mornings is devoted to a rapid survey of simple computer architecture. This serves as a general introduction to the course and more specifically preceeds 'Microprocessor system components' a

An example of an M.Sc. course at UMIST 237

	MODULE ONE MICROCOMPUTER ENGINEERING		MODULE TWO DIGITAL DESIGN			MODULE THREE SOFTWARE ENGINEERING		MODULE FOUR ADVANCED TOPICS
	Monday		Tuesday		Wednesday	Thursday		Friday
9.15 10.05 10.15 11.05	Introductory and revision material $	Microprocessor system components $	Introduction to logic design $	Logic design $	Alternative periods for full-time students to do :- Laboratory, tutorials, design study.	PASCAL laboratory		Finite maths. Silicon I.C. technologies
11.15 12.05	Structure of microelectronic systems I $		Analog/digital techniques & interfacing		Seminars, visits, etc.		$	Digital modelling (option) or Image processing (option)
1.30 2.20	Introductory assembly language programming					PASCAL	$	
2.30 3.20 3.30 4.20	Laboratory & tutorials		Laboratory, tutorials & design study			Introduction to systems software	Software Engineering I $	Laboratory & tutorials
	$ Basis of January Tests for full-time candidates. The Monday module is taken by Communication Engineering and Power Electronics M.Sc. Courses. $ Individual lecture courses taken by Integrated Circuit System Design M.Sc. Course.							

FIG. 1   M.E.D.E. Course, Term One.

	MODULE ONE MICROCOMPUTER ENGINEERING	MODULE TWO DIGITAL DESIGN		MODULE THREE SOFTWARE ENGINEERING		MODULE FOUR ADVANCED TOPICS
	Monday	Tuesday	Wednesday	Thursday		Friday
9.15 10.05	Structure of microelectronic systems II $	Microprogramming $	Alternative for full-time MEDE students to do :- Laboratory, tutorials, design study.	Assembly Language programming	Software Engineering II $	Distributed computing (option) Coding theory (option)
10.15 11.05	System specification design & implementation tools $					
11.15 12.05	Microprocessor applications	Logic testing and simulation (option) $	Feasibility study etc.			Digital signal processing (option)
1.30 2.20	Microprocessor laboratory & tutorials	Fault-tolerant computing (option)		Assembler laboratory		Simulation (option)
2.30 3.20		Laboratory, design study & tutorials				Laboratory, feasibility study, etc.
3.30 4.20						
	The Monday module is taken by Communication Engineering and Power Electronics M.Sc. $ Individual lecture courses taken by Integrated Circuit System Design M.Sc. Course.					

FIG. 2   M.E.D.E. Course, Term Two.

TABLE 2

Day	Title	Participating Department(s)
Monday	Microcomputer Engineering	E.E. & E./Computation Depts.
Tuesday	Digital Design	E.E. & E. Dept.
Thursday	Software Engineering	Computation Dept.
Friday	Advanced Topics	E.E. & E./Computation Depts.

topic lectured in the following five weeks. The emphasis here is on integrated circuit components themselves. Accordingly, items such as memory chips, peripheral handling devices, signal processing chips, DMA chips and the like, receive attention. This is in contrast to the lectures in the series 'Structure of microelectronic systems I' where the architectures of 8-bit microprocessors are described. Emphasis is placed on the 8080/8085 microprocessors which are extensively used in the laboratory classes. For the same reason an introductory course in 'Assembly language programming' based on the Intel 8085 microcomputer is given during the earlier part of the term to support the Monday afternoon laboratory in microcomputer engineering. An outline of the laboratory activity associated with the overall course is given in Table 3. A laboratory course book is provided for each module and distributed to every student to permit preparation in advance of laboratory sessions.

The themes introduced in the first term are continued in the second, as will be seen from Fig. 2. The series 'Structure of microelectronic systems II' extends the ideas of computer architecture developed in 'Systems I' while the overall design process is considered in 'System specification design and implementation tools', which outlines in general terms the tools and techniques which have been developed for the specification, design and implementation of computer-based products.

The Monday module is completed by a series of lectures on 'Microprocessor applications'.

The content of this module has been discussed in some detail to stress that every attempt has been made to provide a package which is intellectually coherent and in which lectures and laboratory activity in microcomputer engineering have been provided in an integrated manner. This approach is particularly important for the part-time student on the M.E.D.E. course and equally vital for students on related M.Sc. courses such as Communication Engineering, for which the other modules of the M.E.D.E. course will not be available. In order to make the arrangements clear to students, the course handbook provides timetables, detailed syllabuses, and reading lists for every module.

Tuesday module: Digital Design
As with the Monday module, this one commences with an element of revision through the lectures entitled 'An introduction to logic design'. These take

TABLE 3 *Laboratory Work.*

Module	No. of hours	Term
1 *Microcomputer Engineering*		
Introduction to Intel 8085 assembly language	5	1
Introductory microcomputer experiments — SDK 85	6	1
Advanced microcomputer experiments — SBC 80/05	6	2
Use of microprocessor development environment	6	2
Use of hardware test aids	3	2
Tutorials	5	1 & 2
2 *Digital Design*		
Introductory digital experiments	6	1
Design study	9	1 & 2
Presentation of design study results	3	2
PLA/ROM based machine design	6	2
Use of a logic simulator	6	2
Microprogramming	6	2
Tutorials	4	1 & 2
3 *Software Engineering*		
PASCAL project	30	1
Assembly language programming	10	2
4 *Advanced Topics*		
Finite mathematics	3	1
Silicon I.C. technologies	3	1
Digital modelling or image processing	3	1
and *two* of the following according to option choice:		
Distributed computing		
Coding theory	6	2
Digital signal processing	(total)	
Simulation		
Tutorials	5	1 & 2
Feasibility study	15	2

place during the first five weeks and are followed by a systematic 'Logic design' course. Since all lecture courses reflect a personal approach by the lecturer, the emphasis in recent years has been on design via the approach of the state machine. These lectures are complemented by a series on 'Analog and digital techniques' with special reference to A/D and D/A converters.

In the second term considerable attention is devoted to microprogramming. After discussing the general concepts of microprogramming, the lectures continue with consideration of microinstruction design and implementation, microprogrammable bit-slice devices and the implementation of a simple computer control unit. Finally, special-purpose processors, such as emulators and high-level language interpreters, together with development aids such as meta-assemblers and simulators are discussed.

The second term also provides a first opportunity for options with a choice between 'Fault-tolerant computing' and 'Logic testing and simulation'. Here, two staff members are able to draw on recent research experience to give the students some flavour of the latest developments in their chosen fields. The provision of such options is regarded as a very important feature of the M.E.D.E. Course. With changes in staff, option topics can be rearranged at comparatively short notice. With this module appropriate laboratory work is provided, see Table 3. A small design exercise is mounted in the Lent term. This is complementary to the PASCAL project of the Michaelmas term which forms an important part of the Thursday module as below.

Thursday module: Software Engineering
This module is provided by colleagues in the Department of Computation. In the Michaelmas term, the morning is devoted to programming practice through the medium of the high-level language PASCAL which is rapidly gaining popularity in computation circles, while a formal one-hour lecture is delivered in the afternoon. PASCAL has been chosen as the language for the M.E.D.E. students because it is felt that it facilitates the design of both the data structures required for the computer solution of problems and the algorithms which operate on the data. It supports the step-wise solution of problems and results in programs which are easy to read and understand. The compiler checks extensively that data are used correctly and thus automatically finds many program errors which, in other languages, could only be revealed by testing. Indeed, PASCAL is a useful tool for the design of algorithms even when a compiler is not available and the final program has to be implemented in some other language. A PASCAL project carried out in the second half of the term is designed to bring out these features. The PASCAL work carried out by the students, together with the logic design project and some of the formal laboratory work is assessed by the staff and is available to the examiners at their meeting following the examinations held in May — but more of this later.

During the Michaelmas term the PASCAL work is complemented by 'Introduction to systems software' and 'Software engineering I'. The first course deals systematically with operating systems, compilers, assemblers, linkers and locaters, together with debugging aids and text processors. The second course concentrates on the problems encountered in designing and implementing software systems such as complexity, correctness and robustness.

In the Lent term 'Software engineering II' carries the subject further by building on the earlier PASCAL software engineering courses. It relates to program maintainance, portability and verification. High-level concepts for expressing concurrency are also introduced to supplement the sequential concepts dealt with in the first term. This course is complemented by 'Data structures' which provides a structured programming approach to data handling. Accordingly, the course discusses arrays and their implementation, sets and their use, dynamic data structures, linear linked lists and pointers and store management.

The software engineering module concludes with a course on 'assembly language programming' centred round the Intel 8085 microprocessor. This course is complemented by an assembly language laboratory where students have an opportunity of putting into practice the topics discussed in the formal lectures.

Friday module: Advanced Topics

With this module, options play an important role. However, two compulsory topics are provided in the Autumn term. The first is 'Finite mathematics and its applications'. Topics considered include groups, rings and fields.

The development of the finite algebraic structures of groups, rings and fields is followed by a study of polynomials with coefficients from a finite or Galois field. Methods for constructing $GF(q)$ for q both prime and composite are investigated. This algebra forms the basis of many applications in digital systems and digital communications and the course includes coverage of two of these. Linear sequential circuits and, in particular, the autonomous feedback shift register are studied from the point of view of its sequences and their applications. The algebra also provides a mode of description for multiple-valued switching circuits. These applications give insights into other areas of digital processes and possible future developments. The course is also intended to provide the foundation for the main application of this work which is exploited in the coding theory option of term two.

The second compulsory topic is 'Silicon integrated circuit technologies'. The objective of this course is to illustrate how silicon wafer processing and fabrication techniques are used to produce the various types of bipolar and MOS silicon integrated circuits which are currently available. This course provides appropriate background material to add point to the various I.C. logic design criteria introduced by other lecturers. It also serves as a useful introduction for those who will be involved with the next generation of I.C. devices. The lecturer is able to draw on considerable personal practical experience together with further background knowledge derived from his strong role as course organiser and participant in the related M.Sc. course 'Integrated Circuit System Design'.

The remainder of the lecture content of this module in Term I and for the whole of Term II relates to optional courses. In view of space considerations a listing of titles alone is possible:

Digital modelling
Image processing
Coding theory
Digital signal processing
Simulation

Table 3 shows that supporting laboratory work is provided during the afternoon practical session.

In each case the topic relates to the special research interests of the staff members involved. In all, eight options are provided and students are encouraged to concentrate on the four topics in which they will be examined in May.

DISSERTATION FEASIBILITY STUDY

Towards the end of the Michaelmas term students are asked to complete a questionnaire which relates to their dissertation topic. They are asked to give details of their previous background including industrial experience, current interests and expectations as to their first job on qualifying. They are also asked to specify four particular topics in order of preference chosen from a list of projects provided by staff members.

The staff, for their part, are encouraged to list a wide range of topics, typically several projects for each staff member. These reflect current interests, work in progress, joint projects with industry etc. Staff are also encouraged to attend a briefing session at which they can speak for a few minutes about their work. This session is of particular value in the case of staff from the two participating departments who do not otherwise encounter the students during the Michaelmas term. Indeed staff from other departments of the Institute are keen to have students for dissertation work so that each year some students carry out projects in the Departments of Mechanical Engineering, Analytical Sciences. Textile Technology and the Control Systems Centre. Others work in the Medical Electronics Unit associated with the Department of Electrical Engineering and Electronics.

At the beginning of the Lent term a small subcommittee of staff from the two participating departments carry out the exercise of matching staff requests and students' wishes. With the help of the questionnaires, the briefing session and a certain amount of informal discussion, this potentially hazardous exercise has become relatively painless. Suffice it to say that for the session 1981–2, seventeen students out of twenty eight received their first choice of project, while most of the remainder achieved their second choice. Perhaps, most significant of all, there were no protests at the allocation from either staff or students. Course organisers reading this account will be well aware of the sensitive nature of project allocation and the problems which frequently follow in its wake.

During the Lent term, students carry out a limited feasibility study as to their project. This is presented to supervisors before Easter and receives a mark which is available to the examiners at their May meeting. The expectation is that the feasibility study establishes the objectives and the method of work for the project, sets up time scales, permits the ordering of special items in good time and ensures that some background reading has been carried out. The intention is that students are thus able to proceed with their project immediately the examination results are announced in late May.

STUDENT ASSESSMENT

The first formal assessment comes at the beginning of the Lent term when two short examinations, each of two-hours duration are held. The topics under consideration are the revision material already mentioned, together with tests on programming in PASCAL and at assembly level. The results are communicated to the students by their personal tutors who are thus able to discuss the students' progress in some detail. While the results are not included with the formal examinations held in May, they form part of the overall student assessment. As such the 'January tests' are particularly valuable.

As already indicated, the course involves a limited amount of assessed course work which is taken into consideration when the examiners consider the results of the written papers in May. The assessed work includes: the PASCAL project; the logic design project; assessed laboratory work; the feasibility study in respect of the project; the January tests.

For each topic the students are made aware of their progress and can take suitable action if progress is proving inadequate.

May examinations

Four formal examinations, each of three-hours duration, are held at the beginning of the Summer Term. Each examination corresponds to a module of the course. Thus any part-time student can take the examination or examinations relevant to the module(s) currently being studied. The results are considered at the Examiners Meeting at which all staff associated with the course are present. One feature of the system is the presence of two staff from departments not associated with the course. These staff are expected to 'moderate' the results should the Board prove unduly harsh or rather too lenient. Since moderators are present at all the M.Sc. examiners meetings throughout the Faculty, the expectation is that a uniform standard will be achieved for all M.Sc. courses. In view of the wide variety of such courses, especially since some involve non-technical subjects such as Management, this expectation may not always be attained.

Perhaps, more important, is the external examiner who is appointed for a three-year period and who acts as the final arbiter at the examiners meeting. However, the principal role of the external examiner is to assess the examination questions both for difficulty and for possible ambiguity. Thereafter the external examiner scrutinises a selection of the student scripts to determine if the marking has been carried out in a conscientious manner. The external examiner may also require oral examination of certain candidates just prior to the examiners meeting. The external examiner is normally a senior academic from another university and has usually acted in the same capacity elsewhere. As a result, a body of opinion as to relevant standards is built up throughout the country. While the process is quite informal, it is nevertheless effective.

Candidates either pass the examination at M.Sc level, or at Diploma level, or they fail. In the majority of cases they pass and proceed to their dissertation project. Those who pass at Diploma level carry out a short project which is assessed internally, and then depart. With the dissertation project, work continues throughout the summer and concludes with the presentation of a thesis by mid-October. The thesis is considered by the supervisor, acting now as internal examiner, and by an external examiner. In some cases the external examiner in respect of the course work takes on this additional burden. The examiners prepare a brief written report on the work of the candidate which is presented to the Postgraduate Committee with a recommendation as to Pass, Pass with Minor Changes, Resubmit after further work, or Fail. The normal expectation is that most candidates will fall into the first or second category. An oral examination is not customary. This is deliberate policy, since it is felt that an oral examination would give the impression to staff and student alike that a very large piece of work is required for the dissertation thesis. Indeed, the omission of an oral examination serves to protect the status of the M.Sc. by research and more specifically the Ph.D degree[1], for which an oral examination is mandatory.

THE FUTURE OF M.Sc COURSES

Taught M.Sc. courses originated in the USA where they had become very popular by the early 1960s. During the decade 1960–70 such courses appeared in the UK and by the end of the decade were well-established. The courses were a response to rapidly-expanding bodies of knowledge and a demand from a large number of potential students to have quick access to such knowledge. Under these circumstances the provision of M.Sc. Courses seemed to be most effective.

However as the writer has already pointed out[1], the mounting of such courses involves very considerable staff effort as to contact hours for lectures, tutorials and laboratory activity. Further commitment as to student recruitment, timetable arrangements and the provision of laboratory facilities is also necessary. Over the years, M.Sc. courses have proliferated in the UK to the extent that many attract only a few students, sometimes of limited ability. In these circumstances the continuation of such courses becomes questionable. The provision of new M.Sc. courses requires most careful thought especially since, in many countries, including the UK, the resources provided for universities are being reduced continuously through limitation in governmental funding.

Nevertheless, such courses may still prove very valuable in certain circumstances. For example with the M.Sc. course discussed in this paper, students receive an integrated training in microprocessor engineering, with emphasis on both the software and hardware aspects of the subject. Many of the students are young engineers from the U.K. who are seeking to extend their existing knowledge, while others are retraining from related disciplines. One or two students are seconded from U.K. Polytechnics or Technical Colleges for retraining in mid-career. The overseas students include several who will return to their parent universities with further qualifications. On their return their new-found knowledges will assist in their teaching at undergraduate and postgraduate level. Clearly such a process is very desirable.

A well-organised course in which staff lecture with enthusiasm on their own particular speciality gives a chance to try out material which will shortly be included as a final-year undergraduate option and, before too long, become part of the regular undergraduate programme. What must be avoided is an M.Sc. course which is mounted in an attempt to be fashionable or to recruit more graduate students. All engineering educators know of such courses and are well aware of their fundamental limitations. As a result of cutbacks in resources, the days of such courses are numbered. It is important that well-established and worthwile courses do not perish also.

ACKNOWLEDGEMENTS

The author wishes to thank his colleagues in the Department of Electrical Engineering and Electronics at UMIST for the discussions which have helped to shape the views expressed in this paper. As organiser for the M.Sc. Course under consideration he is grateful to colleagues for their continued support with the teaching and administrative work.

Successive generations of students on the course have assisted in the formation of views developed in this article. They are thanked sincerely for their contribution.

REFERENCES

[1] Hartley, M. G., 'An Example of an M.Sc. Course — Digital Electronics at UMIST', *Int. J. Elect. Eng'g Educ.*, **17**, pp. 19–32, (1980).
[2] Hartley, M. G., Editorial, *Int. J. Elect. Eng'g Educ.*, **18**. p. 195 (1981).
[3] Hicks, P. J., 'A practical approach to digital integrated circuit design using uncommitted logic arrays', *this volume*, p. 38.

Note

The M.Sc. Course Handbook, 1981–82, *Microprocessor Engineering and Digital Electronics*, is available on application from the MEDE Course Organiser, Dr. M. G. Hartley, E. E. & E. Dept, UMIST, PO Box 88, Manchester M60 1QD, UK